PEACE BY DESIGN

We live in an era in which designers will make an essential and critical contribution to the health and success of humanity. Design can promote healing in healthcare environments, contribute to good mental health, reduce gun violence, and positively impact health and racial equity, all of which contribute to providing a more peaceful world.

The primary focus of this book is to inspire young designers, academics, and practitioners to achieve their maximum societal contribution. It also supports experienced designers seeking reaffirmation of their social goals. To provide a foundation, the first chapter discusses the definition of design and design thinking and evidence regarding the direct and indirect contributions of design to peace. The subsequent chapters address peace endeavors at six scales of the physical environment: sustainable and equitable design, landscape architecture, architecture, interior design, industrial design, and graphic design. Additionally, nine short cameos are provided by contributors from various disciplines, who provide their favorite examples of "peace projects." Peace can be manifested at multiple levels: world-wide, neighborhood and community, familial, or individual, and the various authors discuss portions from this spectrum. They broadly endorse disciplinary entanglements as a means of addressing societal and sustainability challenges and celebrate the impact of collaboration.

This book is essential reading for students and practitioners representing all fields of design, including graphic design, industrial design, interior design, architecture, landscape architecture, and urban design.

Mardelle McCuskey Shepley, BA, MArch, MA, DArch, is a professor in the Department of Human Centered Design, a member of the graduate faculty in the Department of Architecture at Cornell University, and the Academic Director of the Cornell Institute for Healthy Futures. A fellow in the American Institute of Architects and the American College of Healthcare Architects (ACHA), she has LEED AP, WELL AP, and EDAC credentials. Dr. Shepley has published multiple peer-reviewed journal papers and authored/co-authored six books, the most recent of which are: *Design for Critical Care* (2009), *Health Facility Evaluation for Design Practitioners* (2010), *Design for Pediatric and Neonatal Critical Care* (2014), and *Design for Mental & Behavioral Health* (2017). Mardelle has worked full- and part-time in professional practice for 25 years. Committed to designing and conducting research on healthy and healing environments, she received a Lifetime Achievement Award from ACHA.

PEACE BY DESIGN

Edited by Mardelle McCuskey Shepley

Routledge
Taylor & Francis Group
LONDON AND NEW YORK

Designed cover image: Naomi A Sachs, PhD, MLA, EDAC. Naomi is an assistant professor in the Department of Plant Science and Landscape Architecture at the University of Maryland and founding director of the Therapeutic Landscapes Network.

Font: "Adrianne Text designed by Marconi Lima and licensed through Adobe Fonts."

First published 2025
by Routledge
4 Park Square, Milton Park, Abingdon, Oxon OX14 4RN

and by Routledge
605 Third Avenue, New York, NY 10158

Routledge is an imprint of the Taylor & Francis Group, an informa business

British Library Cataloguing-in-Publication Data
A catalogue record for this book is available from the British Library

ISBN: 978-1-032-61101-3 (hbk)
ISBN: 978-1-032-61041-2 (pbk)
ISBN: 978-1-003-46201-9 (ebk)

DOI: 10.4324/9781003462019

Typeset in Galliard
by Apex CoVantage, LLC

CONTENTS

CONTRIBUTORS

Chapters

Rebecca L. Ames with bachelor's and master's degrees in psychology from Tulane University, a minor in architectural studies, and six years of creative industry experience, Rebecca's ability to center creative problem-solving around the human experience has enabled her to traverse disciplines throughout her career, from working directly with youth in clinical settings to developing strategic approaches for healthcare design. She has presented in a variety of venues and published multiple peer-reviewed papers on the topics of trauma-informed design, participatory design research, and color in healthcare environments. Currently, Rebecca is a doctoral student at Cornell University in the Department of Human-Centered Design, majoring in human behavior and design. Her research focuses on developing a comprehensive approach to applying trauma-informed care principles to the design of mental and behavioral health facilities for adolescents to support long-term resiliency.

Rosalyn Cama FASID, NCIDQ, EDAC, holds a bachelor of science degree with distinction in interior design and textiles from the University of Connecticut. She is President and Principal Interior Designer of CAMA, Incorporated in New Haven, Connecticut. CAMA is a health design studio and lab that focuses on designing a better life indoors where one can flourish. Established in 1983, the CAMA team has worked on numerous projects throughout the United States. In 2016 she established Cama Collection, LLC, an evidence-based healthcare product line that partners with several industry manufacturers. A fellow in the American Society of Interior Designers (ASID), she served as ASID's 24th president and is Chair Emeritus of the board for the Center for Health Design and Chair of its Advisory Board. Her awards include an honorary doctorate in fine arts from the New York School of Interior Design, the ASID Designer of Distinction Award, Nightingale Product Awards and the Center for Health Design's Russ Coile Lifetime Achievement Award. Rosalyn has multiple publications, including the book *Evidence-Based Healthcare Design*.

Renata M. Leitão is an assistant professor in the Department of Human Centered Design at Cornell University. She is a graphic designer and social justice–focused design researcher specializing in projects engaging Indigenous and marginalized communities and focusing on the intersection of critical design, participatory design, communication design and design anthropology. She holds a PhD in environmental design, a MASc. in design and complexity (Université de Montréal), and a BA in graphic design (Mackenzie University, Brazil). She was awarded a postdoctoral fellowship by the Social Sciences and Humanities Research Council of Canada. She previously taught in the graphic design program at OCAD University. Raised in a multiracial family, Dr. Leitão is committed to building bridges between diverse worldviews and highlighting the ways of knowing and perspectives of those at the margins. She launched, as co-chair, the DRS Pluriversal Design Special Interest Group, an international research network that aims to highlight multiple perspectives in design, especially from those commonly oppressed by and excluded from mainstream design narratives.

Erin K. Peavey AIA, is Health and Well-being Design Leader at HKS. Erin's passion for creating environments that support health, happiness, and well-being drives her evidence-based, user-centered approach to design. She is an industry scholar at Cornell's Institute for Healthy Futures and served as adjunct faculty at The Pratt Institute and New York School of Interior Design. Erin is a widely sought-after speaker, author, and mentor. Erin has been awarded by the American Institute of Architects, Building Construction + Design, Engineering News Record, and the Academy of Architecture for Health for her work. Her work on design's role in fostering mental and social well-being has been featured on BBC; TEDx; Psychology Today; SXSW; KERA/NPR; and her own podcast, Shared Space.

Troy Savage is the Strategic Projects and Innovation Manager for Mazzetti, an engineering and sustainability consultancy. He holds a BSE in civil and environmental engineering (international affairs, finance, computing application minors) from Princeton University and master's degrees in environmental science and divinity from Yale University. Troy's passion for people and process fuels his approach to helping individuals, organizations and industries innovate. Troy leads and facilitates Reimagining Workshops, coordinates support of health and innovation assistance projects through the International Federation of Healthcare Engineering WHO Techné and helps drive the US healthcare industry's decarbonization efforts. As an ordained minister, Troy shepherded a nascent, reconciliation-focused church community for nearly two years as it simultaneously emerged from the COVID-19 pandemic and reckoned with an unexpected leadership transition. In 2022 he was selected for the U40 (Healthcare Design's Best Under 40) list by the American Institute of Architects, Academy of Architecture for Health.

Mardelle McCuskey Shepley BA, MArch, MA, DArch, is a professor in the Department of Human Centered Design, a member of the graduate faculty in the Department of Architecture at Cornell University and the Academic Director of the Cornell Institute for Healthy Futures. A fellow in the American Institute of Architects and the American College of Healthcare Architects (ACHA), she has LEED AP, WELL AP and EDAC credentials. Dr. Shepley has published multiple peer-reviewed journal papers and authored/co-authored six books, the most recent of which are: *Design for Critical Care* (2009), *Health Facility Evaluation for Design Practitioners (2010)*, *Design for Pediatric and Neonatal Critical Care* (2014), and *Design for Mental & Behavioral Health (2017)*. Mardelle has worked full and part time in professional practice for 25 years. Committed to designing and conducting research on healthy and healing environments, she received a Lifetime Achievement Award from ACHA.

Daniel M. Winterbottom RLA, FASLA, received a bachelor of fine arts from Tufts University and a master of landscape architecture from the Harvard Graduate School of Design and is a professor of landscape architecture at the University of Washington. His firm, Winterbottom Design Inc., focuses on healing/restorative gardens. His research interests include the landscape as a cultural expression, ecological urban design and the role of restorative/healing landscapes. He co-authored *Healing Gardens* by Timber Press in 2015 and has contributed to books on sustainable design, community gardens, therapeutic landscapes, and community service learning. He developed a participatory design/build program in 1995; he and his students work with communities to design and build projects that address community social and ecological concerns. He has completed projects in Seattle, New York City, Bedford Hills New York, Mexico, Guatemala, Bosnia/Herzegovina, and Croatia. He received EDRA's Great Places Award in 2010 and 2019 and numerous ASLA awards and was inducted as an ASLA fellow in 2011.

JungKyoon (Jay) Yoon is an assistant professor in the Department of Human Centered Design at Cornell University where he leads Meta Design & Technology. Dr. Yoon investigates how products can be systematically designed to enrich users' momentary as well as long-term experiences by means of emotions, building on knowledge and methods from human-centered design, positive psychology and persuasive technology. Jay has an educational background in industrial design and computer science (BA and BE from Handong University, South Korea). He previously worked as an interaction designer at Naver Corporation, where he supported the development of an Internet search service, social media and product simulations. He earned his MSc from the Design for Interaction program in industrial design engineering at Delft University of Technology, the Netherlands, where he also received his PhD with a focus on emotion-driven design.

Essays

Evangelia Chrysikou PhD, RIBA, is Associate Professor at the Bartlett School of Sustainable Construction, Program Director of the MSc Healthcare Facilities and a medical architect. She is Vice-President of the Urban Health Section (EUPHA) and ULI Life Sciences and Healthcare Council Leadership Committee Member. Evangelia authored the national guidelines for mental health facilities in the community for Greece on behalf of the European Union and led several international projects on health and the built environment. Dr. Chrysikou authored the books: *Architecture for Psychiatric Environments* and *Therapeutic Spaces* and *The Social Invisibility of Mental Health Facilities.*

Taft Cleveland BArch, MArch, MUP, AICP, Assoc AIA, AORN, NOMA, is an associate and health equity planner at SmithGroup and Co-Founder of the Health Equity Design Collaborative. As an alumnus of Florida A&M School of Architecture+Engineering Technology, he attributes his achievements to the exemplary support and sense of community fostered among Black architects and designers throughout the HBCU diaspora. Taft's specialties include architecture, urban design, master planning, health care planning, graphic presentations, and creative problem solving.

Marie Elf RN, PhD, is a professor of nursing at Dalarna University with a PhD in healthcare architecture from Chalmers University of Technology. Her research explores the critical link between the built environment and its impact on individuals with stroke and older adults, emphasizing how physical spaces influence health and well-being. Beyond her academic interests, which span nursing, architecture, and healthcare management, she is committed to interdisciplinary approaches. She actively bridges academic knowledge with practical applications, aiming to improve the lived experiences of vulnerable groups through research and collaborative efforts across various sectors.

Denise Nicole Green BS (with honors in fashion design), MS (textiles), PhD (socio-cultural anthropology), is an associate professor of fashion design and management and Director of Graduate Studies in Fiber Science and Apparel Design at Cornell University, where she also directs the Cornell Fashion + Textile Collection. Green is an award-winning curator, documentary filmmaker, fashion designer, and writer. She combines ethnographic methods, creative design, and archival research to explore how fashion transforms individuals, communities, and ecologies.

Connie Lin BArch, MA, LEED AP, WELL AP, EDAC, is both an architect and researcher. She is currently a fellow at Perkins&Will. Her education includes a graduate degree from Cornell University and a postgraduate program in neuroscience applied architectural design at the Università Iuav di Venezia. Connie has also served as an adjunct professor, guest lecturer and studio juror for several universities. Her research addresses the spatial implementation of evidence-based design focusing on biophilic design, environmental psychology and mental health outcomes.

Lorraine E. Maxwell BA, MCRP, PHD, professor emerita at Cornell University, is an environmental psychologist with extensive experience in facility planning and programing. Her research interests focus on children and are related to the ways in which the physical environment impacts their development, behavior and well-being. She is highly committed to community service. She was a tutor in the afterschool program at a local elementary school and a participant in the Big Brother/Big Sister program for five years. She currently serves as the chair of Grants Committee for the City Federation of Women's Opportunity of Ithaca (CFWO) and is an emeritus member of the EDRA Board (Environmental Design Research Association). She served on the advisory board for the 21st Century School Fund in Washington, DC.

Arsenio T. Rodrigues AIA, NCARB, PHD, LEED AP, EDAC, is a registered architect in India and the United States. He is the founding Director of the Bowling Green State University School of the Built Environment. Previously, Arsenio served as Director and Assistant Dean of the School of Architecture at Prairie View A&M University, an historically Black college. In 2022, he was named Architect of the Year in Toledo in recognition of his contributions to the profession.

Tammy Thompson EdD, MArch, NCARB, CPXP, EDAC, is a registered architect, interior designer and health care leader whose career has focused on improving patient experiences, with a focus on health equity. She is a hospital administrator, and she serves on the engineering faculty of North Carolina A&T State University.

Renato Troncon Educated as a philosopher specializing in German aesthetics and phenomenology, Renato Troncon is a senior professor of aesthetics, design epistemologies and ethics at the University of Trento. He is also the Co-founder and Coordinator of the Design Research Lab at the University of Trento, as well as a research associate at the Istituto Italiano di Studi Germanici in Rome, Italy. His teaching and research interests encompass aesthetics, philosophical anthropology, service and community design, biophilic design and regionalism. A highly cited writer, he has authored or edited approximately 15 books. Additionally, Renato is the President of Iprase, the institute responsible for teacher training in the Province of Trento. He is actively involved in the field of cultural policies in both Italy and Germany.

ACKNOWLEDGMENTS

I would like to thank Sue Watson for significant contributions to the detailed review of two chapters; Allison DeDominick for the graphic rendering of several figures; Fran Ford, senior editor and publisher, and Hannah Studd, editorial assistant at Routledge, for their guidance during the publication process; and Rebecca L. Ames for her contributions as a research assistant.

I am very grateful for the contributions of the chapter and essay authors, all of whom are courageous proponents of peace.

M.M.S.

PREFACE

Mardelle McCuskey Shepley

This is a great moment to be a designer. We have the potential to be the ushers of sustainability regarding the physical environment, masters of visualization in the service of society and harbingers of a social perspective that acknowledges the role of design in our interactions, learning and healing. We are becoming more inspired in our aspirations, more caring in our endeavors and more integral to social progress. Now, perhaps more than ever, design is a social art.

We have not always stood at the forefront of social transformation. Deserved or not, designers have had a reputation for being driven by their egos à la Ayn Rand and eschewing accountability. The Pruitt-Igoe housing project serves an example of design misdirection augmented by financial crises and racial discrimination (Bristol, 1991) (see Figure 0.1). Even "starchitects" such as Corbusier have stumbled when the local culture and urban behavior were not thoroughly incorporated in design decisions, as was the case in Chandigarh, India (Kalia, 1985). Are we delivering what we are promising and ethically directed to do? Unfortunately, learning how to assess our success regarding meeting human needs isn't part of many design school curricula.

Peace by Design was inspired by predecessors such as Victor Papanek. Papanek's books, including *Design for the Real World and Social Change* (1972) and *Design for Human Scale* (1983), promote the principled rejuvenation of the profession. In addition to aesthetics, Papanek advocates for the functionality of a product regarding the individual and society at large. His mandate reminds us that we also need to be accountable, avail ourselves of research

FIGURE 0.1 Pruitt-Igoe low-income housing was occupied in 1954 and demolished between 1972 and 1976.

Source: "Pruitt-Igoe-collapses_editado" (www.flickr.com/photos/69365082@N05/7472172108) by Renato Saboya (2012) (www.flickr.com/photos/renatosaboya/) is licensed under CC BY 2.0 (https://creativecommons.org/licenses/by/2.0/).

and engage in multidisciplinary collaboration (Papanek & Fuller, 1972). Beyond collaboration, a transdisciplinary integration of expertise is necessary to achieve our goals, and designers have demonstrated their preparation to reinvent themselves. Our research indicates that individuals trained as architects translate their design education into strong problem-solving skills in the fields of politics, cuisine, finance, dance, journalism and medicine.

A photograph I took in April of 2020 (see Figure 0.2) shows a fence with a quotation from Dorothy Day (1897–1980), "Beauty will save us." Day, a social radical and suffragist, whom I profoundly admire, was referencing Fyodor Dostoyevsky's quotation from *The Idiot*, "beauty will save the world" (Hennessy, 2017). The wooden gateway on which these words were painted served as a canvas for local graffiti artists and marked the head of the trail lying in the railway bed for the Cayuga-Susquehanna train that once meandered from Ithaca to Pennsylvania. The message greeted those who were seeking nature to distract themselves from the COVID-19 pandemic. As I absorbed its meaning through my designer lens, it struck me to the core.

FIGURE 0.2 "Beauty will save us" graffiti art.

This epiphany was one of several that spawned this book. Most healthcare design researchers study healthcare settings or translate research into healthcare design guidelines, and I wondered whether there was an additional way to use our skills to address larger-scale social issues. Another call to action had occurred to me a couple of years earlier. The frequency of gun incidents prompted my colleagues and me to publish a literature review on the impact of greenspace on violent crime (see Chapter 1), followed by a meta-analysis of greenspace/gun violence data and literature review on community collaborative research associated with outdoor spaces.

The issue of gun violence was an initial catalyst for this book. George Floyd's death threw fuel into the motivational fire, and the impetus behind the collaboration grew.

Hence the mission of this book. Promoting environmental healing, reducing gun violence and accepting responsibility for health and social/racial equity have one thing in common: they can contribute to a more peaceful world. As designers, how might we use our training to address these and related issues to contribute to the promotion of human well-being? The authors of *Peace by Design* invite you to read their stories and share your own with us and others.

References

Bristol, K. G. (1991). The Pruitt-Igoe myth. *Journal of Architectural Education*, *44*(3), 163–171. https//doi.org/10.2307/1425266

Hennessy, K. (2017). *Dorothy Day: The world will be saved by beauty: An intimate portrait of my grandmother*. Simon & Schuster.

Kalia, R. (1985). Chandigarh: A planned city. *Habitat International*, *9*(3/4), 135–150.

Papanek, B. (1983). *Design for human scale*. Van NostrandReinhold.

Papanek, V., & Fuller, R. B. (1972). *Design for the real world: Human ecology and social change*. Pantheon Books.

Saboya, R. (2012). Pruitt-Igoe-collapses_editado [Photograph]. *Flickr*. https://www.flickr.com/photos/69365082@N05/7472172108

SCALES OF PEACE AND AUTHORS' CONTRIBUTIONS

Peace-seeking endeavors manifest at multiple scales, ranging from engaging in planetary climate action to manipulating a smooth stone in your palm. With this breadth in mind, the various chapter and essay authors in this book represent a spectrum of designers. The chapter authors bring their multidisciplinary knowledge via six design disciplines: design process, landscape architecture, architecture/environmental psychology, interior design, industrial design, and graphic design. The essay authors provide vignettes of specific experiences. Each author describes the way in which their specialization supports health, wellbeing, sustainability, and social justice. Combined, they demonstrate the power of design to contribute to peace.

Chapter 1: How Beauty Will Save Us; Chapter 8: The Architecture of Peace; Editor *(Mardelle McCuskey Shepley).* In Chapter 1, Mardelle McCuskey Shepley provides definitions of design and design thinking and discusses indirect and direct ways in which design can contribute to peace. Topics include design justice and sustainable design. Chapter 8 integrates the content of the other contributors and reiterates the mission of peace by design. Mardelle is an architect, author, academic, and design researcher focusing on the impact of the physical environment on physical and mental health.

Chapter 2: WeDesign: Participatory Innovation as a Basis for Reimagining Possibilities (*Troy Savage*). Design is more likely to be successful when those impacted can share in the process, especially if those impacted can participate in design from the fullness of their being. Where many are impacted or where problems are intransigent, this may be a new source of innovative energy that can lead to peace. Troy Savage addresses this theory and practice of "WeDesign" in Chapter 2. He brings his multidisciplinary credentials in engineering, policy, environmental science, and divinity, as well as his work across industries (from consulting to academic to religious), to the conversation.

Chapter 3: Cultivating Peace *(Daniel M. Winterbottom).* In Chapter 3, Daniel M. Winterbottom explores the positive contribution of nature, gardening, and landscape to trauma recovery and its role in achieving spiritual, psychological, and physical tranquility. Nature interactions have been used as a refuge from the ordeal of war and to support post-war conciliation. Based on his experience in participatory design, he discusses strategies to address social and ecological concerns. Daniel M. Winterbottom is a landscape architect, researcher, and author with expertise in sustainable design, community gardens, therapeutic landscapes, and community service learning.

Chapter 4: Design for Peace Through a Trauma-Informed Perspective *(Erin K. Peavey and Rebecca L. Ames).* Interaction with the physical environment impacts mental health. Individuals who experience trauma are particularly susceptible. Trauma-informed design principles in support of trauma-informed approaches are discussed in Chapter 4. The authors also address peace as a deeply personal experience, a state of inner harmony, and a contentment that fosters our sense of belonging and our ability to form meaningful relationships

with others. Erin K. Peavey is an architect, speaker, writer, and researcher who focuses on the gap between health design research and design practice. Trained in architecture and psychology, she advocates for mental health via the physical environment. Rebecca L. Ames is a researcher in the field of environmental psychology, who is currently studying the application of trauma-informed care principles to the design of the built environment to support long-term resilience and growth.

Chapter 5: Peace for a Life Indoors *(Rosalyn Cama).* Interior designers were among the earliest champions of evidence-based design. In Chapter 5, interior designer Rosalyn Cama discusses how the design of the interior environment can elicit a sense of peace as well as evoke interpersonal harmony. Cama, who has had a 40-year career in healthcare interior design and is the founding principal at CAMA, has been an early champion of the evidence-based design movement. Focusing on providing a better life indoors, she has published and spoken widely and designed numerous projects throughout the US.

Chapter 6: Everyday Products as a Means of Uplifting Inner Peace, Meaning, and Happiness. *(JungKyoon (Jay) Yoon).* In Chapter 7, Jay Yoon explores notions of "happiness" and "well-being," revealing that design engendered happiness is largely dependent upon users' activities and the broader environments in which a product is used. JungKyoon (Jay) Yoon is an industrial designer, academic, and researcher. Yoon investigates how products can be designed to enrich users' momentary and long-term experiences by means of emotions and methods from human-centered design, positive psychology, and persuasive technology.

Chapter 7: Empowerment by Design (*Renata M. Leitão*). In Chapter 8, Renata M. Leitão examines the connection between design and the empowerment of Indigenous and marginalized local communities. While believing in the positive force of design, she reminds us of the danger of well-intentioned but ill-informed interventions, noting the importance of allowing local stakeholders to define and act upon their own projects. Leitão is a graphic designer and social justice-focused design researcher specializing in projects engaging Indigenous and marginalized communities. Raised in a multiracial family, Dr. Leitão is committed to decolonizing design research and education.

Cradled in the middle of this book are short essays by designers from a variety of disciplines, nationalities, and career trajectories. Contributions have been provided by Renato Troncon, Tammy Thompson, Taft Cleveland, Lorraine E. Maxwell, Connie Lin, Arsenio T. Rodrigues, Marie Elf, Evangelia Chrysikou, and Denise Nicole Green. Their short papers are called "cameo essays" as a way of celebrating their unique and, in some cases, rarely published positions regarding advocacy for peace by design.

What do all these authors have in common? Although their methods and disciplines are disparate, the chapter and essay authors of this book believe in the role of design products and processes in supporting human well-being and personal and interpersonal peace. They represent the fields of philosophy, engineering, urban design, landscape architecture, architecture, environmental psychology, interior design, fashion design, industrial design, and graphic design. However, the chapters and essays are not meant to be read by specific design professionals; the content is both common and parallel to the interests of all designers.

The chapter and cameo authors each address the impact of design on peace. Combined, they generate a design and research agenda that should be of interest to those who acknowledge the social responsibility associated with the design professions.

1

HOW BEAUTY WILL SAVE US

Mardelle McCuskey Shepley

Redefining Design and Design Thinking

Design is an innately human activity. Both our ability to adapt and our prowess at creating or selecting an abode would have been critical to survival; therefore, our attention to the environment and the ability to design would have supported our evolutionary progress. "Design predates the professions" (Friedman & Stolterman, 2015, p. vii), or, simply put, "Design is what human beings do" (Thackara, 2005, p. 1). As design is integral to who we are and wellness is a goal of humanity, designers have a significant responsibility to support human health. "Any attempt to separate design, to make it a thing-by-itself, works counter to the inherent value of design as the primary, underlying matrix of life" (Papanek, 1972, p. 268).

According to Friedman and Stolterman (2015), the definition of design has three parts: 1) the impact on the physical world, 2) relevance to human need, and 3) physical manifestation. In the context of this definition, they ask us to consider including the potential of design to expand the context of the problem, the possibility of content beyond the value of the physical product, and the lack of clarity of boundaries between "artifacts, structure and process" (p. ix). The meaning of artifact and structure is somewhat clear, but what is process?

Problem-solving and sense-making contribute to the definition of the design process (Manzini, 2015), but the topic is more complex. Costanza-Chock (2020) notes that design is neither inductive nor deductive, but rather it requires both inference and speculation. Inference, or abductive reasoning, as Costanza-Chock describes it, is the process of attempting to design with incomplete information, while speculation refers to conjecturing about the

DOI: 10.4324/9781003462019-1

future. "Designers imagine images, objects, buildings, and systems that do not yet exist. We propose, predict and advocate for (or, in certain kinds of design, warn against) visions of the future" (p. 15). They go on to say,

> Design thus may be thought of as both a verb and a noun, a universal kind of human activity and a highly professionalized field of practice (or several such fields), a way of manipulating future objects and systems using specialized software and an everyday use of traditional knowledge embedded in indigenous lifeways, a type of work with one's hands and a way of thinking, an art and a science, and more.
>
> *(Costanza-Chock, 2020, p. 16)*

There is some consensus regarding what is meant by ***design***, but the meaning of ***design thinking*** is more nebulous. I once asked a student who had just completed a course titled "design thinking" to define the term, and they were unable to describe it. Brown (2009) refers to design thinking as design with a small "d" to distinguish it from design with a capital "D" or *objet*. As I came to better understand the more contemporary meaning of design thinking – which I now interpret as the process associated with engaging and focusing on users – the definition conflicted with my original interpretation of design thinking, which was derived from cognitive and perceptual psychology and associated with organizing schemata and engaging memory. My prior definition of design thinking represented internalized processes as opposed to the common understanding now, which is emblematic of externalized interactions. Acknowledging both definitions, I suggest that one can engage in design from a cognitive perspective without engaging in design thinking from a service perspective. Perhaps training or native skill in design per se is not required to engage in design thinking?

Brown (2009) notes two goals of design thinking as they relate to service: 1) demonstrating how a human-centered approach offers more solutions and 2) acknowledging how the process may go beyond the skills of a professional designer. Design thinking can be both a preamble to an internalized process and reiteratively integrated into the externalized process. Therefore, there may be a difference between being a professional designer and someone who utilizes the contemporary concept of design thinking. In summary, not all professional designers thoroughly engage in design thinking as a service, and not all design thinking practitioners employ the cognitive processes associated with professional designers.

I was intrigued while listening to Natasha Jen's 2018 lecture, "Design Thinking Is Bullsh*t," because I am careful to attend to individuals who challenge the status quo. I understand Jen to say that designers have been engaging in the process of serving users for a while; hence, the fanfare around the term is not of interest to her. In my career, however, I've noted that commitment to engaging with users/clients comes and goes over time. Jen's derogatory response to design thinking service may actually be an indicator that the idea of human-centered design has officially arrived. I saw a similar capricious response regarding the notion of evidence-based design (EBD). Initially, few understood exactly what it was; then people understood it and considered it a good idea. This affirmation was followed by designers who challenged it, saying that the science wasn't valid or reliable and undermined aesthetics! I knew that research-informed design had 'arrived' when the nay-sayers challenged the approach. Now most people understand that EBD is a humble offering providing some evidence in lieu of none.

Jen claims that design thinking, as a service, was created to help engineers be designers, although some would disagree that this was its sole *raison d'etre*. Brown (2009) states that the expression was first embraced by David Kelley, founder of IDEO. I propose that its roots lay in the mid 20th century via design researchers such as Henry Sanoff, Wolfgang Preiser, Gary Evans, Powell Lawton and Harold Proshansky (to name a few), many of whom were associated with the Environmental Design Research Association. To my knowledge, the contemporary expression "design thinking" had not been coined at that point, although the activities of these trailblazers strongly resemble the hallmark steps (empathize, define, ideate, prototype and test).

Can Design Impact Peace?

In a world where peace seems continually at risk, the charge to contribute to harmony is a responsibility of us all. In the context of the previously mentioned definitions of design and design thinking, we can respond more effectively to the question: Can design impact peace? As the primary creators of discord in the world, humans are theoretically equally capable of supporting peace, or, more explicitly stated, "If we can design our way into difficulty, we can design our way out" (Thackara, 2005, p. 1). Maybe. I have dug a few holes from which I have not been able to extricate myself, and certainly humanity, constrained by misdirects emanating from our core nature, has been floundering since recorded times. But the possibility of doing good exists.

Addressing this question from a historical perspective, the 2010 Netherlands Architecture Institute conference organizers asked its participants: "Is an architecture of peace possible?" (Junne, 2010). Summarizing the event, Junne reported that, following World War I, modernism was intended to reduce future conflict by creating an architecturally classless society, focusing on rationally designed spaces for living and working. This approach faltered but was re-invigorated after World War II and subsequently stimulated by the social activism of the 1960s and 1970s (Junne, 2010). Enthusiasm for socially motivated architecture waned, but, more recently, decades of conflict and the associated urban destruction have generated an interest in designing healing cities (Junne, 2010). This history suggests that architecture can serve as a tool for recovery, but it is unclear whether it can stem conflict *pre hoc*.

Jon Kolko, author of *Wicked Problems: Problems Worth Solving* (2012), suggests that design cannot solve complex, multi-sourced problems (such as conflict/war), but it can mitigate them. He defines wicked problems as social or cultural dilemmas that impede solution due to lack of or conflicting information, the number of persons and perspectives or the cost and interconnectedness of multiple problems (Kolko, 2012),

> most social problems – such as inequality, political instability, death, disease or famine – are wicked. They can't be "fixed." But because of the role of design in developing infrastructure, designers can play a central role in mitigating the negative consequences of wicked problems and positioning the broad trajectory of culture in new and more desirable directions.
>
> (*p. 11*)

Tim Brown, CEO and president of IDEO, is also optimistic about the role of design in abetting progress. In *Change by Design* (2009), he argues that design thinking can contribute to the reconfiguration of societal systems, resurgence of participatory democracy and redesign of cities. As an application of reconfiguring social systems, he offers an IDEO case study involving the redesign of a school system in Peru through multidisciplinary teams. Regarding the resurgence of participatory democracy, he cites the design of better voting machines to enhance our confidence in democracy, and, regarding cities, he notes the role of urban design to adjust to the eventual decline of the automobile. Other large-scale endeavors by IDEO are redesigning the interface with artificial intelligence to humanize technology, redesigning "life and death" through the development of DNA tools and redesigning the fiscal future by helping businesses to understand the circular economy. Clearly these efforts contribute to change, but they may not be core catalysts. IDEO's approach, however, does remind us that we should be continually thinking about designing for the larger context in which a problem is situated.

Cary (2017) attributes power to the impact of design on activities related to healing and provides examples of spaces that enlighten, enhance solidarity and empower, such as the Women's Opportunity Center in Rwanda (see Figures 1.1 and 1.2). Other notable examples of built work intended to support recovery are provided in Charlesworth's book, *Humanitarian Architecture: 15 Stories of Architects Working After Disaster* (2014). Apart from these larger scale interventions, "small actions can have big effects" (Thackara, 2005, p. 14), such as engaging students in work for Habitat for Humanity (see Figure 1.3) or invigorating one's professional design practice by giving children residing in developmental and mental health environments a voice in design decisions (see Figures 1.4 and 1.5).

FIGURES 1.1 AND 1.2 Women's Opportunity Center, Rwanda by Sharon Davis Design.

Sources: Elizabeth Felicella (left image), Sharon Davis Design (right image).

FIGURE 1.3 Architecture students working on housing for Habitat for Humanity in Bryan, TX.

Are these environmental interventions minor gestures, or do they have a ripple effect? According to Cary, to be effective, a designer must have a beginner's mindset, treat clients as partners, build community support, employ local workers, use local materials and measure the impact of their work. Along similar lines, Manzini (2015) writes that to achieve social innovation, a collaboration must occur between trained designers and a world in which everyone has a role in design. However, to create the mindset that promotes these tactics, we need fundamental changes in design education, increased demand for

FIGURE 1.4 A child's drawing, created during design programming for a residential mental health environment, with a stove, game room and stereo and views of the hills.

FIGURE 1.5 The pain of some children, who participated in the exercise, impeded their inclination to collaborate. This young artist, when asked why he had drawn a monster, said that the space was in the mouth of the creature.

social design, the sustenance of alliances with similar values, more funding and more pro bono service contributions (Cary, 2017).

From Cary and Manzini we can confidently assume that, while human transactions are the primary source of wellness and achieving peace, the physical environment can both indirectly and directly support these goals.

Indirect Impacts

Per J.J. Gibson's Affordance Theory (1977), spaces contribute to the actualization of activities. Examples are round tables providing equal stature where negotiations take place or clear physical boundaries where territorial designations are required. As an illustration at the building scale, children in the Camarillo Developmental Center tended to engage in more unsafe behaviors in shared corridors than rooms designated as private or semi-private spaces (Shepley, 1995). Perhaps "good fences make good neighbors?"

However, not all physical environments, regardless of identical configuration, result in the same outcomes. Björkdahl and Kappler (2017) propose that agency (i.e., the ability to promote peace) lies at the intersection of space and place. Place can be geographical, generated by activities or represented in lived experiences. Space is imaginary "referring to the processes of meaning-making and the ideational component of locations" (Björkdahl & Kappler, 2017,

p. 19). They note that places can transition into space by making them relevant to political and social issues:

> Material places, without the counterpart of symbolic, meaningful space, will be limited in their ability to catalyze transformation. Similarly, a purely symbolic idea of space without any material underpinning also remains ideational and lacks sustainability.
>
> *(Björkdahl & Kappler, 2017, p. 144)*

To their point, peace is a transaction between community members or people, or internalized transactions within oneself (Bulan et al., 2017). Hailing back to our previous discussion about placing design in the context of larger objectives, we should couple spatial design with the goals offered by Healy (2018) as strategies for achieving peace (see Figure 1.6).

Many of these goals imply that empathy may be the first step toward peace and human-centered design. Empathy, at both large-scale and intimate levels, is required to achieve a culture of harmony (Murphy, 2022). Kolko (2012)

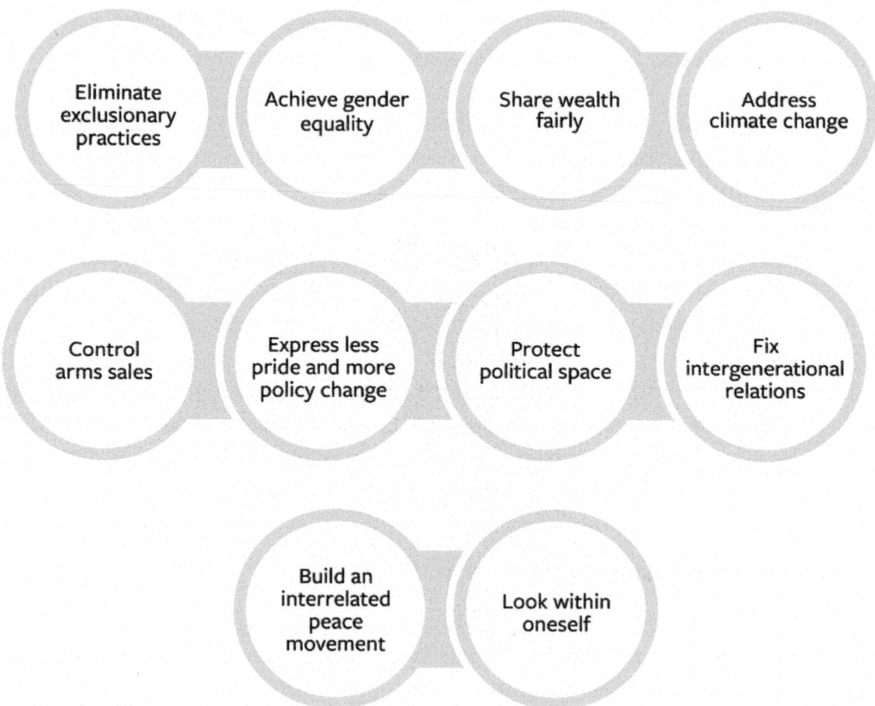

FIGURE 1.6 Healy's goals for achieving peace (graphic based on Healy, 2018).

proposes using research as a means for developing empathy, including contextual inquiry (ethnography with a focus), participatory design and cultural probes (e.g., surveys and photovoice), as well as methods for synthesizing data and creating new designs (e.g., storyboards and journey maps) and bodystorming (conducting research within the environment under consideration). Pastalan's Empathic Model (1977), which is a tool for simulating the physical environmental experience of an older adult, is an excellent example. Empathy, as mentioned previously, is the first step in the contemporary design thinking process.

Direct Impacts

Regarding the direct impact of the physical environment on peace, this happens as a result of either supportive interactions between people and settings or unconscious spatial cues. In the former case, if the environmental experience provides choice and control, the individual is less likely to foment conflict. Regarding unconscious cues, positive distraction has been linked to stress reduction. Positive distractions deflect attention from undesirable circumstances and negative states of mind. Examples are social interaction, music, art and access to nature. Can nature per se encourage more peaceful behavior? The findings on the impact of greenspace on violent crime suggest that this is the case.

An in-depth literature review on greenspace and violent crime had mixed findings regarding the impact of parks per se, likely due to the variations in design and difficulty in funding the development of these spaces in association with pre- and post-studies (Shepley et al., 2019). On the other hand, smaller interventions, such as community gardens, vegetated streets and walkways and, particularly, areas with trees and ground cover, were more likely to be associated with reductions in violent crime (Shepley et al., 2019) (see Figure 1.7).

In summary, the physical environment can both indirectly and directly support peace by creating spaces that afford certain activities and by improving mental health. Expanding upon this, there are two adjacent phenomena that significantly benefit peace goals – design justice and sustainable design.

Design Justice, Sustainable Design and Peace

The provision of design justice will undoubtedly contribute to peace. When resources are distributed equitably, our awareness of our common community is enhanced. As Cary (2017) frames the issue, "Dignity is to design what justice is to law and health is to medicine . . . dignity is about knowing your intrinsic worth and seeing that worth reflected in the places you inhabit" (p. 18). Design justice ranges from the equitable distribution of urban resources to

Type of Greenspace	Reduction in crime	No significant difference	Increase in crime
Parks	Brown (2018) Nitkowski (2017)	Blair et al. (2017) Lee (2013)	Abu-Lughod (2006) McCord et al. (2017)
Community gardens/greening	Heinze et al. (2018) Kondo et al. (2016) Sadler et al. (2017)	Blair (2014) Blair et al. (2017) Gorham et al. (2009)	Wilcox et al. (2013)
Vegetated streets and walkways	Burley (2018) Harris et al. (2018) Harris (2018)	Auchincloss (2019) Locke et al. (2017)	
Trees and ground cover	Gilstad-Hayden et al. (2015) Kondo et al. (2017) Kuo & Sullivan (2001) Snelgrove et al. (2004)	Donahue et al. (2011) Li (2008) Lim (2005) Schusler et al. (2018)	
Undeveloped green areas		Sparks (2011)	

FIGURE 1.7 Research comparing greenspace to frequency of crime, based on Shepley et al. (2019).

the role the physical environment plays regarding food access equity. Design justice is associated with "a more equitable distribution of design's benefits and burdens, meaningful participation in design decisions; and recognition of community-based, Indigenous, and diasporic design traditions, knowledge, and practices" (Costanza-Chock, 2020, p. 23). There are challenges with distributing resources equitably. How do we democratically share assets when there are multiple needs and resources are limited?

Prioritization can be informed by goals and values. The five goals of the Design Justice Network are: 1) *encourage* communities to design themselves, 2) *enable* people to practice from their own knowledge base, 3) *allow* the community to explore its own reality, 4) *state* the problem and possible solutions and 5) *develop* a model of the broader system generating the problem (Escobar, 2017). Regarding values that might be associated with these objectives, Griffin (2022) recommends we embrace acceptance, aspiration, choice, democracy, engagement, fairness, identity, mobility, power, resilience, rights and welfare. Several of these values are contained within or implied by the ten principles of the Design Justice Network (2018), which are summarized in Table 1.1. For those seeking strategies associated with the principles, I have generated potential actions. These recommendations are preliminary and would benefit from the input of readers.

Sustainability tactics are directly linked to design justice, as the lack of access to resources can create conflict. As general guidelines for setting sustainable practices, the Hannover Principles developed by McDonough and Braungard

TABLE 1.1 This table details the principles of the Design Justice Network (2018) and potential actions that can be taken to support these principles.

	Design network principle (Design Justice Network, 2018)	*Potential action (author)*
1	"We use design to sustain, heal and empower our communities, as well as to seek liberation from exploitative and oppressive systems."	Start planning with grassroots users or communities to set goals for sustaining and healing.
2	"We center the voices of those who are directly impacted by the outcomes of the design process."	Formally engage recipients equally in the design process, including with salary compensation.
3	"We prioritize design's impact on the community over the intentions of the designer."	Acknowledge the desires of the recipients as prerequisites leading to proposed goals of the designers.
4	"We view change as emergent from an accountable, accessible, and collaborative process, rather than as a point at the end of a process."	Evaluate completed projects based on the initial design goals set by the designer/recipient team.
5	"We see the role of the designer as a facilitator rather than an expert."	Provide equal status for designers and recipients in the design process.
6	"We believe that everyone is an expert based on their own lived experience, and that we all have unique and brilliant contributions to bring to a design process."	Incorporate contributions of all participants into the design strategy proposal.
7	"We share design knowledge and tools with our communities."	Knowledge is power. All materials used during the design process should be shared with all at each stage of the process.
8	"We work towards sustainable, community-led and -controlled outcomes."	Engage communities in such a way as to generate outcomes that are manageable by or useful to them.
9	"We work towards non-exploitative solutions that reconnect us to the earth and to each other."	Produce design work that acknowledges the larger context in which the project occurs.
10	"Before seeking new design solutions, we look for what is already working at the community level. We honor and uplift traditional, indigenous and local knowledge and practices."	Conduct a pre-evaluation to assess the quality of existing design solutions and utilize the information to inform subsequent designs. Do this under the guidance of community members.

for the 2000 World's Fair are foundational. They share many similarities with those of the Design Justice group, specifically regarding design accountability and the sharing of knowledge. The nine Hannover Principles are:

1 Insist on rights of humanity and nature to co-exist.
2 Recognize interdependence.
3 Respect relationships between spirit and matter.
4 Accept responsibility of consequences of design.
5 Create safe objects of long-term value.
6 Eliminate the concept of waste.
7 Rely on natural energy flows.
8 Understand the limitations of design.
9 Seek constant improvement by sharing knowledge.

(William McDonough Architects, 1999, pp. 8–9)

Sustainability can be couched in the context of the ***pluriverse***, which is described by Escobar as distinct from the ***universe*** as, rather than a singular, focused perspective, it operates with multiple threads and is "ceaselessly in movement, an ever-changing web of inter-relations involving humans and non-humans" (2010, p. 8). The pluriverse involves design at both community and individual levels in the interest of social transition and sustainability and the "Liberation of Mother Earth" is a basic principle and "the most viable historical project that humanity can undertake at present" (Escobar, 2017, p. 204). In summary, the tenets of design justice and sustainable design can stimulate social responsibility. Social responsibility is foundational to policies and environments that support health and wellbeing, which in turn impact personal and interpersonal peace.

The implications of the pluriverse tie back to the intentions of this book in that the various design fields are synchronous at some points in time and asynchronous in others, with frequent shifts in structure, self-awareness and intent. The following chapters and essays demonstrate this phenomenon.

References

Abu-Lughod, R. A. (2006). *Planning for crime reduction: Analysis of social, economic, and physical variables on United States cities* (Publication No. 3229564) [Doctoral dissertation, The University of Texas at Arlington]. ProQuest Dissertations and Theses Global.

Auchincloss, A. H., Michael, Y. L., Kuder, J. F., Shi, J., Khan, S., & Ballester, L. S. (2019). Changes in physical activity after building a greenway in a disadvantaged urban community: A natural experiment. *Preventative Medicine Reports*, *201*, 15. https://doi.org/10.1016/j.pmedr.2019.100941

Björkdahl, A., & Kappler, S. (2017). *Spacebuilding and spatial transformation: Peace, space and place*. Routledge. https://doi.org/10.4324/9781315684529-1

Blair, L. (2014). *Community gardens and crime: Exploring the roles of criminal opportunity and informal social control* (Publication No. 3685404) [Doctoral dissertation, University of Cincinnati]. ProQuest Dissertations and Theses Global.

Blair, L., Wilcox, P., & Eck, J. (2017). Facilities, opportunity, and crime: An exploratory analysis of places in two urban neighborhoods. *Crime Prevention and Community Safety, 19*, 61–81. https://doi.org/10.1057/s41300-016-0011-2

Brown, C. L. (2018). *The built environment and crime: A comparative study of Detroit and Philadelphia* (Publication No. 13419799) [Master's thesis, University of Delaware]. ProQuest Dissertations and Theses Global.

Brown, T. (2009). *Change by design: How design thinking transforms organizations and inspires innovation.* HarperCollins.

Bulan, C., Hashimoto, J., Meyer, K., Ribao, V., Stanich, J., & Wynne, M. (2017). Designing peace: Peace memorial. *Mānoa Horizons, 2*, 178–190.

Burley, B. A. (2018). Green infrastructure and violence: Do new street trees mitigate violent crime? *Healthy Places, 54*, 43–49. https://doi.org/10.1016/j.healthplace.2018.08.015

Cary, J. (2017). *Design for good: A new era of architecture for everyone.* Island Press. https://doi.org/10.5822/978-1-61091-794-0

Charlesworth, E. (2014). *Humanitarian architecture.* Routledge. https://doi.org/10.4324/9781315776545

Costanza-Chock, S. (2020). *Design justice: Community-led practices to build the worlds we need.* MIT Press. https://doi.org/10.7551/mitpress/12255.001.0001

Design Justice Network. (2018). *Design justice network principles.* Retrieved November 25, 2022, from https://designjustice.org/read-the-principles

Donahue, J. D. (2011). *An empirical analysis of the relationships between tree cover, air quality, and crime in urban areas* (Publication No. 1491329) [Master's thesis, Georgetown University]. ProQuest Dissertations and Theses Global.

Escobar, A. (2010). Latin America at a crossroads: Alternative modernizations, post-liberalism, or post-development. *Cultural Studies, 24*(1), 1–65. https://doi.org/10.1080/0952380903424208

Escobar, A. (2017). *Designs for the Pluriverse: Radical independence, autonomy and the making of worlds.* Duke University Press. https://doi.org/10.1215/9780822371816

Friedman, K., & Stolterman, E. (2015). Series forward. In D. Manzini (Ed.), *Design, when everybody designs: An introduction to design for social innovation* (pp. vii–xi). MIT Press. https://doi.org/10.7551/mitpress/9873.003.0001

Gibson, J. J. (1977). The theory of affordances. In R. E. Shaw & J. Bransford (Eds.), *Perceiving, acting, and knowing.* Lawrence Erlbaum Associates. https://doi.org/10.4324/9781315467931

Gilstad-Hayden, K., Wallace, L. R., Carroll-Scott, A., Meyer, S. R., Barbo, S., Murphy-Dunning, C., & Ickovics, J. R. (2015). Research note: Greater tree canopy cover is associated with lower rates of both violent and property crime in New Haven, CT. *Landscape & Urban Planning, 143*, 248–253. https://doi.org/10.1016/j.landurbplan.2015.08.005

Gorham, M. R., Waliczek, T. M., Snelgrove, A., & Zajicek, J. M. (2009). The impact of community gardens on numbers of property crimes in urban Houston. *Horticulture Technology, 19*, 291–296. https://doi.org/10.21273/horttech.19.2.291

Griffin, T. L. (2022). Imagining the just city. In C. E. Smith (Ed.), *Designing peace: Building a better future now* (pp. 36–39). Cooper Hewitt.

Harris, B. (2018). *The invisible walls of the 606: An examination of the relationship between an urban greenway and community change* (Publication No. 10808205) [Doctoral dissertation, Clemson University]. ProQuest Dissertations and Theses Global.

Harris, B., Larson, L., & Ogletree, S. (2018). Different views from the 606: Examining the impacts of an urban greenway on crime in Chicago. *Environment and Behavior, 50*, 56–85. https://doi.org/10.1177/0013916517690197

Healy, H. (2018, September 18). 10 steps to world peace. *The Internationalist*. Retrieved January 1, 2023, from https://newint.org/features/2018/09/18/10-steps-world-peace

Heinze, J. E., Krusky-Morey, A., Vagi, K. J., Reischl, T. M., Franzen, S., Pruett, N. K., Cunningham, R. M., & Zimmerman, M. A. (2018). Busy streets theory: The effects of community-engaged greening on violence. *American Journal of Community Psychology, 62*, 101–109. https://doi.org/10.1002/ajcp.12270

Jen, N. (2018, August 22). Design thinking is bullsh*t [Video]. *YouTube*. https://www.youtube.com/watch?v=V8gjDsW3lsY

Junne, G. (2010). Designing peace: Bricks and mortar of reconciliation. *The Broker, 20*, 30–33.

Kolko, J. (2012). *Wicked problems: Problems worth solving*. Austin Center for Design.

Kondo, M., Hohl, B., Han, S., & Branas, C. (2016). Effects of greening and community reuse of vacant lots on crime. *Urban Studies, 53*, 3279–3295. https://doi.org/10.1177/0042098015608058

Kondo, M. C., South, E. C., Branas, C. C., Richmond, T. S., & Wiebe, D. J. (2017). The association between urban tree cover and gun assault: A case-control and case-crossover study. *American Journal of Epidemiology, 186*, 289–296. https://doi.org/10.1093/aje/kwx096

Kuo, F. E., & Sullivan, W. C. (2001). Environment and crime in the inner city—Does vegetation reduce crime? *Environment & Behavior, 33*, 343. https://doi.org/10.1177/00139160121973025

Lim, J. (2005). *The moderating effects of vegetation on human violent behavior caused by environmental stressors* (Publication No. 1430668) [Master's thesis, The University of Texas at Arlington]. ProQuest Dissertations and Theses Global.

Locke, D. H., Han, S., Kondo, M. C., Murphy-Dunning, C., & Cox, M. (2017). Did community greening reduce crime? Evidence from New Haven, CT, 1996–2007. *Landscape & Urban Planning, 161*, 72–79. https://doi.org/10.1016/j.landurbplan.2017.01.006

Manzini, D. (2015). *Design, when everybody designs: An introduction to design for social innovation*. MIT Press. https://doi.org/10.7551/mitpress/9873.001.0001

McCord, E. S., & Houser, K. (2017). Neighborhood parks, evidence of guardianship, and crime in two diverse US cities. *Security Journal, 30*, 807–824. doi:10.1057/sj.2015.11

Murphy, M. (2022). An architecture of peace. In C. E. Smith (Ed.), *Designing peace: Building a better future now* (pp. 26–33). Cooper Hewitt.

Nitkowski, J. (2017). *Physical environment and crime in Milwaukee neighborhoods* (Publication No. 10607501) [Master's thesis, University of Wisconsin-Milwaukee]. ProQuest Dissertations and Theses Global.

Papanek, V. (1972). *Design for the real world: Human ecology and social change*. Pantheon Books.

Pastalan, L. A. (1977). The empathic model a methodological bridge between research and design. *Journal of Architectural Education, 31*(1), 14–15. https://doi.org/10.2307/1424529

Sadler, R. C., Pizarro, J., Turchan, B., Gasteyer, S. P., & McGarrell, E. F. (2017). Exploring the spatial-temporal relationships between a community greening program and neighborhood rates of crime. *Applied Geography, 83*, 13–26. https://doi.org/10.1016/j.apgeog.2017.03.017

Schusler, T., Weiss, L., Treering, D., & Balderama, E. (2018). Research note: Examining the association between tree canopy, parks and crime in Chicago. *Landscape and Urban Planning, 170*, 309–313. https://doi.org/10.1016/j.landurbplan.2017.07.012

Shepley, M. (1995). The location of behavioral incidents in a children's psychiatric facility. *Children's Environments*, 352–361.

Shepley, M., Sachs, N., Sadatsafavi, H., Fournier, C., & Peditto, K. (2019). The impact of green space on violent crime in urban environments: An evidence synthesis. *International Journal of Environmental Research and Public Health, 16*(24), 5119. https://doi.org/10.3390/ijerph16245119

Snelgrove, A. G., Michael, J. H., Waliczek, T. M., & Zajicek, J. M. (2004). Urban greening and criminal behavior: A geographic information system perspective. *Horticulture Technology, 14*, 48–51. https://doi.org/10.21273/horttech.14.1.0048

Sparks, C. S. (2011). Violent crime in San Antonio, Texas: An application of spatial epidemiological. *Methods in Spatial Spatio-Temporal Epidemiology, 2*, 301–309. https://doi.org/10.1016/j.sstc.2011.10.001

Thackara, J. (2005). *In the bubble: Designing in a complex world*. MIT Press. https://doi.org/10.7551/mitpress/3702.001.0001

Wilcox, P., Quisenberry, N., & Jones, S. (2003). The built environment and community crime risk interpretation. *Journal of Research on Crime and Delinquency, 40*, 322–345. https://doi.org/10.1177/0022427803253801

William McDonough Architects. (1999). *The Hannover principles: Design for sustainability: For the City of Hannover, Germany, EXPO 2000, the world's fair* (5th ed.). W. McDonough Architects.

2

WEDESIGN

Participatory Innovation as a Basis for Reimagining Possibilities

Troy Savage

As a designer/engineer and minister whose work involves teaching people how to work together to create innovative solutions to problems, I see the tremendous potential for designers to foster peace. There are many seemingly intractable problems in the world today, among them climate change, health equity, and reconciliation. All these need solutions for the world to truly live in peace. These problems are not typically viewed from a design perspective, but they should be. Solutions to them need to be designed; that is, they require beauty, should be based in substance, and should be expressed through culture towards a particular end.

We live in a paradox. How can such an inequitable world exist when, arguably, most people express a desire for a more equitable world? Why has this expressed and lived desire of people been held back?

World problems are impacted by the internal life of individuals and by their outward manifestation. As such, they are often very complex problems that present themselves in ways that make us believe that there are no design solutions. Part of me (in reality, most of me) used to think our most difficult challenges were impossible to solve on this side of eternity – even our most endowed *imagineers* could not produce viable solutions. Admittedly, that part of me still exists, but it is slowly being replaced by an understanding of what is possible if we reimagine and redesign our possibilities.

This chapter weaves together an approach to thinking about people, discusses a partial model of this method, and examines Reimagining Workshops as a precursor of the proposed model.

DOI: 10.4324/9781003462019-2

Artists and Designers

When we think of revealed beauty expressed through culture, we imagine the artist. The exemplar of this is the fiction writer, speaking to the universal truth because they speak to and through the particularity of their culture. Writers like Toni Morrison and Derek Walcott are control theorists of culture, finding the equations and feedback loops that make culture observable and controllable; they interpret and create. In this model, the writer, by dint of her peculiar genius, understands herself, her culture, and her words and then writes in ways that illuminate and transcend them – exposing universal truths because of her very deepest understanding of ultra-local truth.

We see the best of design in that same light – the designer as an artist. Yet the designer also crafts that expression towards a particular end. Designers are both scientists and engineers. The best design captures something spectacular – what architect Christopher Alexander might call "the quality without a name," and makes it observable. (See Alexander, 1979.) That which is beautifully designed is both applied science and high art. Its source is the genius of the artist.

There is another model for the role of artists and designers. What if the great writer is not an interpreter of culture but instead a participant in its observability? What if writers are not creators but instead revealers? What if design is also construed as an act of revealing instead of creating? In this version, the foundational questions are what is revealed. But what are the mechanisms for this revealing?

This chapter focuses on seeing design as the exploration of individuals and cultures at their roots and effectively collaborating to achieve solutions to our most pressing challenges from multiple vantage points. It is intentional designing by synchronizing diverse perspectives in ways that, when combined, achieve something whole.

Whose Design Is It Anyway?

Philadelphia, Pennsylvania, USA, is the home of Constitution Hall, the original seat of the US government, and the Liberty Bell. I was visiting the city with a native Philadelphian, someone who had lived her entire life in the city. As we looked up at the majestic, tall buildings, she marveled and shared the thought, "They can build these incredible buildings. Why can't we? Why could they build these buildings, and we couldn't?"

I knew her question was important then, but now, almost a decade later, I've finally recognized the profundity of her statement. Why would someone who lived in a city her entire life not feel connected to the building of its infrastructure? Why see it as a product of a different world? In an authentic sense, she had not participated in creating the downtown Philadelphia skyscrapers. They existed before she did. But neither had those people with whom she was juxtaposing herself. The buildings were built before their working lifetimes too. For my friend, there was a lack of personal connection to the buildings and to anyone who had worked on them. Additionally, there was a lack of cultural connection. It didn't feel like her people – however they were defined – made the skyscrapers; it felt like someone else's product. In other words, it didn't proceed out of her individual or cultural experience.

I've had the inverse conversation with people, taking pride of ownership of something to which they had not contributed personally, but this was the first time I explicitly heard this version expressed. What makes someone feel disconnected from the architectural majesty of a city, while others feel connected and treat it as part of their inheritance and legacy – or to any experience, for that matter? The reasons are likely many, significantly complicated, challenging, and mostly beyond the scope of this chapter.

One could argue that it is my friend's steadfast participation in the economy of Philadelphia that made these innovations viable.

Within the confines of a small geographic area, even over time, we are all inextricably bound. We marvel at what we find most beautiful, awe-inspiring, inciting, and thought-provoking, and, in some cases, we may even seek to own it. Some may even proclaim, "my people built this." However, we don't often say, "This captures something of the eternal, and so it is mine." We rarely say, "We designed it! It is ours!"

But why not? What if it were our design? What if, instead of marveling at another's work, we tangibly participated in its creation? What if the design belonged to all of us instead of one or two of us?

The design and its implementation might improve because it would be responsive to many instead of just those who had direct agency in it. Carefully conceived, it could enhance equity by ensuring our voices were heard in the process. Carefully executed, it might be transformational, creating new breakthroughs and new possibilities.

I am not talking about simply "design by committee." I am talking about bringing the fullness of people to a design process that allows that fullness to be directed to the right challenges and design approaches to both more possibilities on the front end and deeper impact on the back end. One of the many tasks a collaborative process can achieve is to include the full experience of all of the people impacted.

How can it be everyone's design? How can this work? First, we should conceive of design as choosing the form, function, and aesthetic to achieve desired ends. Everything is designed – skyscrapers, teapots, books, and organizations. It can be our design if: (a) we fully participate in a portion of the design process, and (b) the outcome of that process yields something reflective of and responsive to our participation. Participating starts with consideration of the problem space – defining the specifics of the challenge for which a design solution is sought. It then moves on to the solution space that reflects the challenge identified in the problem space.

Fullness of Who We Are

No model can encompass an individual completely, but often developing a stylized concept is helpful in applying our "fullness" to design challenges. Our fullness is composed of (a) our experiences; (b) our culture, or what we believe our culture to be; and (c) our worldview. These three are intertwined. Our experiences are shaped by both what we believe our culture to be and our worldviews. We also have experiences that impact what we know to be true and how we experience the world. These experiences may be colored by parts of our identity. Some parts of people's culture, worldview, and experience are helpful, and some parts are unhelpful towards a particular end. We often, implicitly and explicitly, ascribe value judgements to culture. Instead, once a desired direction is defined, the question becomes what is most helpful to reach an outcome.

Worldview is created by our culture, our faith traditions, our experiences, and the universe of ideas that surround us. In our most hopeful moments, when we can imagine beyond the lived experience and the barriers the world has placed upon us, we may see an idyllic world. It is a world where we are loved and supported and can live our worldview without denigrating the experience of others.

Solving intractable problems involves understanding and activating the fullness of individual participants both within the problem space (carefully defining the problem we are trying to solve specifically while leaving degrees of freedom for the solution) and within the solution space of a design process. This is most critical when we feel our hopes are unrealistic, that they can't happen, that this isn't the world we live in, and that we need to deal with reality. Yet the worldviews of both hopeful moments and realistic moments are important because they underlie how we operate. For challenges rooted in how people operate and interact, understanding individuals' worldviews is essential.

We all live in multiple worlds with experience in applying different worldviews – or perhaps different components of their worldviews – depending on their situations. While true for everyone, it is tangibly visible for two types of people: (a) those rooted in a faith tradition and (b) those who are marginalized.

For people of faith, holding to a worldview based on an understanding transcendent to sensory experience is part of religious practice. Those who are marginalized often live in one world for "things official" and live in a different world for their daily bread. Living in two or more worlds and developing the capacity to maintain two worldviews is an essential moniker of resilience. Some individuals are more experienced in switching between worlds and expressing or living in the paradox between different worldviews because it is necessary for their material survival. This experience is invaluable for design because this latent resilience is a source for new design possibility. Yet when we engage with people who live in two worlds (including ourselves) and who, therefore, might have two worldviews, which perspective surfaces in interactions? If these individuals who have learned to deal with two worldviews feel free to allow both to surface, could this better inform the design solutions to challenging world problems?

Assisting people to access their lived experience and culture is essential because these are at the root of innovation. Yet most social processes and design processes ask the marginalized for their worldview and cultural experiences drawn from commonly held centrist beliefs, instead of that which could be derived from the margins. This is important enough to say again. That which is at the center implicitly forces what's at the margins to conform to the center. For the marginalized, learning to speak the language of the center is required for existence. The result is that we rarely get fullness.

Creativity and innovation are often seen as creating something new, but often they are not. Instead, they usually involve applying something old to a unique situation or circumstance. Interaction with diverse people and ideas is a method of providing learning to new situations beyond current realms of understanding. Yet that interaction is assumed from the center instead of from the margins. That's why participation in a design process does not always fully meet the intentions. The goal is not to "look" diverse but to be diverse. What is posited here is that we need to co-discover and activate people's fullness and then constructively allow this to be incorporated into design problems and solutions.

The vital advantage of this approach is that it compensates for the circumstance that one or two people cannot hold the sum experience of everyone. Providing opportunities for people to participate fully allows the sum of what we learn from our beautiful, diverse cultures, mindsets, experiences, and individuality to be directed towards tangible and intangible people and earth-enhancing benefits. For the biggest problems we have, we need the exponential impact of our multiple experiences directed towards these challenges.

What Is the Role of Design?

Design impacts everything we do. How can we think about designing for people, health, and the environment? How can we put people at the center of what we're doing as designers? How might we build communities that exist in a designed world and design this world for peace? What are the ideas and worldviews that help with this process?

A colleague and I created a workshop for architects and designers on "choice architecture," a term attributed to Thaler and Sunstein (2009). One of the premises is that everything is designed, and there are choices (conscious or unconscious) behind all associated decisions. Yet the danger of being a designer is that we can become consumed with the idea of design. We might think that design is about our creativity instead of a broader story. Simply stated, design is ingenuity applied to things to achieve an impact. Everything is designed, although everything may not be perceived as designed.

If everything is designed, what is at the center of the design, and, more importantly, what is at the periphery of the design? Often, because we don't consider the margins we don't acknowledge that the negative space that surrounds the positive space must also be designed. However, we assign moral judgment to what is a direct and predictable outcome of this unconsidered space. I suspect my "City of Brotherly Love" friend saw the skyscrapers as beautiful but, because she did not participate in the design, did not feel connected to them. Relatedly, we face challenges, such as health and environmental equity, the social determinants of health, and reconciliation, that require designed solutions and for which the impact is paramount for everyone, especially those in the margins who may feel disconnected.

Choosing the problems on which to focus design solutions is important. Accessing the fullness of lived experience within this problem space can be powerful as a source of innovation because it impacts what we decide to address. Greater effort can be placed into the design of things that have a more significant impact. One of the reasons that the exploration of outer space has had so much impact on Earth is because it has inspired people to solve significant problems, learn more about the universe, and create new technologies and new understandings. Generations of bright people have dreamed of solving the problems of the universe. However, we still have vast problems that require collective human ingenuity. What if we innovatively explored the heart of the issues that impact the masses from the perspective of the masses? What might this do to reorient a world towards reconciliation, sustainability, and equity?

What if we thought about a new approach to designing for world peace based on accessing the fullness of many people when defining challenges and defining solutions? To do so, we'd need to enhance everyone's ability to participate by bringing together a diverse group of people and using a process that allows their energies to be effectively directed and harnessed to promote beautiful solutions. This is the essence of what I interchangeably call WeDesign or Reimagining.

It is difficult, if not impossible, to operate in fullness in groups. It involves clear role definition and engaging in complex social dynamics. It can be easy not to try. Yet anything closer to fullness will likely lead to better results.

Instead of not even trying, take the mindset that everything and anything is possible and that what is inside of people, especially when built upon, might be incredibly useful for the world if it can be expressed. We want to bring to the surface everybody's full contribution to define what is most beneficial and use that definition as a basis for directing people's total contributions. We do this without judgment. There is objectivity around the possibility, efficiency, and efficacy of what we have inside us when applied to specific places, contexts, and situations.

How we think about design is critical for the most pressing challenges we face. Activating the fullness of individuals, including their individuality, culture, worldview, and faith, and defining the right challenges based on an activated group of designers is helpful. Only then can value judgments between people regarding their culture and experience be avoided. Only then can an analysis of what might work best in solving a challenge, especially for those who are most impacted by the challenge, occur. In bringing people together to design, "my design" can become "our design," providing voice and agency to those affected by the design.

This is WeDesign. To make WeDesign happen well and to build a WeDesign-infused innovation pathway, we need to consider a model to unleash these possibilities.

The E Models for WeDesign

For pressing challenges involving shared impact and shared responsibility, including systematic racism, climate change and sustainability, health equity, and global inequality, and in cases where fundamentals of life, beliefs, and lifestyles are implicated, it is helpful to use an innovation platform for designing. The framework presented here may only be ideal in some situations, and, indeed, you should use the mindsets and frameworks best suited for the challenge you are facing. The goal is to bring everyone's entire being to the surface and allow people to work together to direct and design. The "E-models" for co-design presented here are meant to enhance an existing design system. The design system, which will be addressed later in this chapter, is the Reimagining System. We used this system in our Reimagining Workshops to implement a human-centered design and Lean process-infused pathway. (For a definition of Lean design, see Koskela et al., 2002.)

The E-models are not a replacement for other specific design frameworks. Instead, the E-models highlight the importance of engagement and other essential aspects of working on a problem to uncover potential solutions. There are eight components of the E-model: engage, empathize, explore, express, enmesh, and experiment, plus two additional aspects that are not part of the design process but are infused within it: equity and energy (see Figure 2.1).

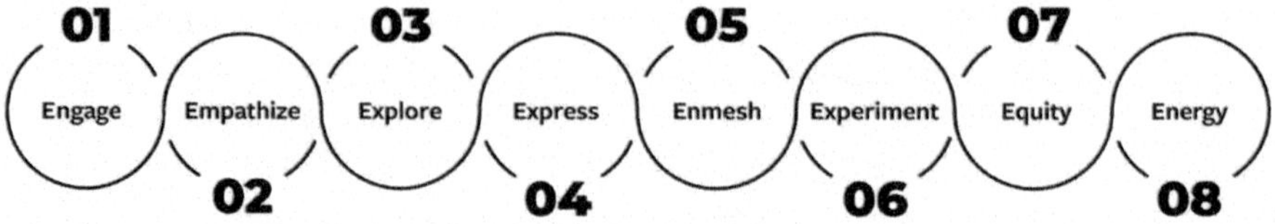

FIGURE 2.1 Eight components of the E-model.

1. Engage

The engage stage is a critical portion of this process, as it prepares everyone participating to live in the moment and bring as much of themselves and their worldview to the table as possible. In practice, this is done by allowing people to share their comfort and discomfort in a collegial environment. This is done by warming up the participants.

To warm up, slowly bring people to the edge of their comfort zones (comfort interaction, in communication) in a space/place/topic where they do not perceive anything to be at stake. This is analogous to a warm-up for physical activity, where muscles are brought to this place in preparation for the game. An example may be helpful. You might ask people to:

1 Name their favorite chocolate bar out loud,
2 Work with their team to create ideas for a new chocolate bar,
3 Imagine a culture exactly as their own but where chocolate bars didn't exist and work with their team to determine a way to promote chocolate bars in that culture,
4 Individually identify the key attributes of the culture to which they imagined promoting the chocolate bar,
5 Identify the key attributes to their team, and
6 Discuss if their chosen promotion might work if one of those key attributes were changed.

Activities should be designed to engage the complete person, including worldviews, cultures, experiences, and individuality. Ideally, engagement allows people to move to the other side of what may hold them back from fully participating in the experience.

There is much to be achieved through engagement. Engagement needs to break down barriers between people. Since this process is meant to allow people to fully engage themselves and others, it is vital that people slowly develop the ability to trust and work with one another. The process should allow people to understand one another's ideals, experiences, and culture. Engagement needs to open people to identifying problems and exploring solutions without disengagement. It needs to create, within individuals and a group, the capacity for resilience in moments of frustration, lack of energy, and lack of trust. We must develop the possibility of not shutting down in the face of someone else's humanity. Ideally, engagement would allow all this to occur. In my experience, even with rudimentary attempts, this process makes a huge difference.

Engagement comes at some risk. Some people may react to engagement as "a waste of time" that could be spent on "the real content." For some, especially those historically left out of processes (either explicitly or implicitly from signals that suggest that their fullness was not accepted), engagement may be questioned as unauthentic or "too little too late." It may take time for people to become comfortable with engagement. Yet the reward is the possibility of entering a design space with the fullness of individual and cultural knowledge present and ready to identify and solve the most pressing challenges.

One caveat: it's essential to consider the difference between engagement and equilibrium. Part of the purpose of engagement is to reduce unhelpful power dynamics that develop in groups. Power dynamics are unhelpful, as they hamper or prohibit people from fully participating within a group. Often, people and groups can reach an equilibrium – a helpful dynamic – because it feels safe and allows people to move forward quickly. However, the goal is not to reach any equilibrium; the goal is to engage. The equilibrium is one where everyone is fully participating, since that is what is necessary for solving pressing challenges. Engagement, therefore, does not necessarily equate to reaching a working equilibrium and instead may be more akin to finding rewards and freedom amid discomfort.

2. Empathize

Empathy, a part of the human-centered design process, is essential. Empathy is helpful not just because it involves listening but also because it involves explicitly identifying which voices and expertise to center on. We center on those with lived experience of a particular challenge. In situations where everyone has varied experience, how do you decide which voices to focus on in empathy? The answer is to focus on those with the most experience in the system/situation and those most negatively impacted. Those who are most impacted by the status quo will also be most impacted by a proposed solution.

Empathy is an attempt to become one with another's reality. Empathy requires understanding the bounds and extents of self while making room to experience another's essence. In its ideal state, this experience is from the other's vantage point, not your own.

Systems create benefits for some and detriments for others. What is it about the combination of person and system that leads to experience? This is the locus of understanding we want to gain through empathy. Engagement is vital in creating empathy. In this formulation, everyone is in designer/discovery mode. Full engagement permits the eventual right focus for empathy.

3. Explore

Exploration happens after engagement, so that people can live entirely in their humanity while exploring possibilities with one another. Exploration becomes possible when people move beyond social hierarchies (real or perceived). This exploration space is an opportunity for deep learning with one another. It requires an open mindset, ideally achieved after engagement and empathy. In some senses, exploration is simply a natural result of empathy.

4. Express

Participants must clearly express the problem about which they are co-designing from the perspective of those most negatively impacted by the current situation. Expression of ideas is the locus of the challenge to be solved. This is akin to the Define stage in human-centered design or design thinking theory. (See Deepa, 2020.) What did we learn from empathy that we might address? Where is the thrust of the energy that we need to assert to make significant progress?

5. Enmesh

This can be interpreted as "de-enmeshing" or taking off "mental shackles" while remaining true to your worldview. A source of innovation is seeing things differently, which is why diversity (of all types) is essential in innovation. We want to retain our worldviews while removing the mental shackles and blind spots associated with that worldview. The point isn't to give up your experiences and expertise but to make us more able to use them beneficially to solve challenges. To do this, we need to both appreciate our worldview and consciously hold it less tightly than we are perhaps used to in hopes of learning and building from others. This is the point at which deep learning happens and where ideas flow, since you can learn from another's experiences and vantage points and allow this learning to build upon your own experiences and ideas.

6. Experiment

Experimentation is critical to all innovation. In human-centered design and design thinking, experimentation is the prototype and test phase. The difference is that experimentation explicitly highlights the exploratory nature of the endeavor. However, these experiments might initially feel incongruent because they may be outside of our typical experience.

What is important here is the mindset we use for experimentation. Experimentation feels different when it involves not just one idea or another but, instead, ideas that might touch the substance of our being. We are embracing the fullness of self and yet not holding too tightly to self. We are experimenting with the possibility of specific outcomes and avoiding judgment about the outcomes. In experimenting, we honor the possibility that another way of doing something might be beneficial. It involves creating simple, quick ways to test key assumptions and then testing those assumptions. Based on the learnings from these tests, we continue to seek new opportunities.

In addition, there are two additional Es that are not part of the design process but should be infused within it:

7. Equity

Equity should be at the center of the design process. This is best accomplished when a highly diverse group of co-designers is brought to the table. Therefore, those curating a design opportunity should intentionally gather a diverse group of people to participate – enough diversity so that people can feel "we helped create that." Equity is an output of the process, and although it is not a filter or a step (though, if necessary, it could be included as a tollgate between steps), enhancing equity is a goal of the process.

8. Energy

The overall energy of the group and its implications for individual energy should be recognized and managed. Understanding where people are individually and collectively on a personal energy cycle can help things move more smoothly. It is okay to be frustrated or unsure; we'll come to a beneficial end in a robust innovation experience. It's okay to have moments of doubt about other participants in the group or about whether they can see you. The process itself is meant to understand and manage energy and take advantage of opportunities where people are energized to do something.

By tracking, understanding, and building the energy of a working group, opportunity is provided.

There are moments when energy is high, and people are motivated to do work, believe in what is possible, and that they might come together to do something. There are also moments where energy is low, where people are frustrated and cannot see beyond a particular sticking point. It is helpful to provide opportunities for forward progress in both the high-energy and low-energy moments.

High-energy moments provide a sense of what is possible in the future and what can be achieved together. Low-energy moments may be equally important because they lead to high-energy moments. How? Low-energy moments track with moments where people tend to consider only their own viewpoint and experiences. There is a reversion to what is comfortable. Breaking through the low-energy moments provides a tangible example that there is new possibility in being open to the full contribution of other designers. Therefore, even in low-energy moments, it is crucial that the energy be directed in ways that are productive. You don't want the energy to get so low that forward progress feels impossible. So, interactions should be structured to allow forward progress.

These E-model additions to a design process are meant to be applied within distributed teams of six to eight people to create the possibility of the fullness of culture and experience being applied to the design process. Teams are small enough to engage meaningfully, and multiple teams work in parallel such that a group of any size can be working simultaneously on the same topic. Using the Es provides an opportunity to match different areas of expertise and direct them towards understanding the physical and cultural realities that define our problems and the design solutions to those problems.

Implications for the Role of Design

At the time of this writing, I have only scratched the surface of applying these concepts and techniques. Projects in which I've been involved range from considering how artificial intelligence can be used in the neonatal intensive care unit to enhance patient-centered care to changing paradigms regarding designing space in response to a pandemic. I see the possibility of harnessing rivers of uninhibited inner beauty in productive and beneficial ways. There is a wellspring of newness and possibility, especially in designing for peace through WeDesign. This is not simply co-designing but instead a framework that recognizes the innate possibilities for beauty, innovation, and ingenuity vested in our experiences and culture.

The E-models are insights into how these rivers may be accessed. They are not meant to stand on their own but are intended to enhance existing design processes. As an example of how the models can be applied, they are discussed in this chapter in the context of the Reimagining Workshops.

An Example – Reimagining Workshops

Reimagining Workshops use human-centered design and Lean process improvement–infused techniques to guide small groups in understanding the challenges faced by clinicians, patients, and families in specific healthcare spaces. Launched in 2017, the Reimagining Workshops are an innovation platform at the intersection of clinical design and practice. The first, Reimagining the ED (i.e. the emergency department), was facilitated by DC Design's Durell Coleman, a Stanford-educated human-centered design practitioner and now a dear friend. At the time of this writing, we've hosted several workshops, including Reimagining Childbirth Facilities, Reimagining Critical

Care, Reimagining+ Racial Justice in Healthcare, Reimagining the NICU, Reimagining Behavioral Emergency Environments, Reimagining the OR (i.e. Operating Room), and Reimagining Assisted Living. These workshops bring together clinicians, patients/patient advocates, architects, engineers, code officials, technologists, and others. Participants are placed in small groups of six to ten people. Each group works together on a single challenge chosen by the group. They warm up together and then engage in empathy, focused on the lived experience of the clinicians and patient/patient advocates in the group. They define a challenge they would like to address. Then, in a deviation from the human-centered design process, they spend time creating a simple process map of the challenge identified by the group. Next, they ideate solutions and create simple prototypes of their idea. (See Figure 2.2.) Through cross-group testing, testing with clinicians/or patients, the group's ideas are tested, and then the group reiterates based on the feedback they receive. (See Figure 2.3.) This process repeats a few times. The groups then consider the possibility of code change proposals based on what they experienced as potential barriers or enablers during their work and share their work. (For more details see Shepley et al., 2022.)

I've led hundreds of doctors, nurses, architects, and engineers through this Reimagining process. In doing so, I have come to understand the Reimagining Workshops as an innovation process that might help enhance healthcare facility design and healthcare delivery by creating a space to innovate based on the lived experience of people in clinical spaces. Idea kernels are discovered, and patterns for enhanced healthcare space are revealed. The workshops focus primarily on the intersection of clinical design and practice. Still, groups are not limited in the solutions they explore – the only specification they are given are the workshop topics. Critical current challenges are identified by those with lived experience in the work. We imagine revamping these workshops to include the E-model more intentionally.

For many Reimagining participants, even the fundamental level of engagement we do today has implications beyond the workshop content. Some participants have described these workshops among their most positive experiences. Something is empowering about entering an engagement space where even nascent portions of your being are appreciated, treated as valuable, and built upon to create something new. Especially at Reimagining the NICU and Reimagining Behavioral Emergency Environments, patients/advocates/families were encouraged to share their experiences. This led to new understandings of the challenges faced, and almost all attendees attested to the power and perspective they gained by having people in their group who were empowered to share their stories. It was also crucial that all voices be elevated in the defining, ideating, prototyping, and testing phases, which organically happens when we engage in WeDesign.

FIGURE 2.2 Participants in the Reimaging Assisted Living workshop engaging in process mapping.

FIGURE 2.3 Participants prototyping during the Reimaging Assisted Living workshop.

FIGURE 3.1 Central Park is one of the most revered places in New York City and renowned for its bucolic qualities and sinuous architectural forms. It is a nature-based counterpoint to the surrounding urban bustle, a place of peace and escape.

Source: "Fall in Central Park" (www.flickr.com/photos/49115235@N06/15746400912) by Third Blade Photography (www.flickr.com/photos/thirdblade/) is licensed under CC BY 2.0 (https://creativecommons.org/licenses/by/2.0/).

This belief in nature as a critical component of healing that can stimulate personal feelings of peace can be found amongst some of the earliest societies and is still valued today by many of these same cultures. Current research now scientifically validates these early societies and Olmsted's early intuitions, which were instead based on observation.

Evidence

Biophilia was popularized by Eric Fromm, a psychoanalyst in the 1960s who joined the word bio, meaning life, and philia, meaning friendly towards, to describe a biological drive towards self-preservation. E.O. Wilson, an American biologist and professor at Harvard University, used the same term to describe "the existence of a biologically based, inherent human need to affiliate with life and lifelike processes." Kellert goes on to expand, noting that the "human need for nature is linked not just to the material exploitation of the environment but also to the influence of the natural world on our emotional, cognitive, aesthetic, and even spiritual development" (Kellert & Wilson, 1995, p. 40). The term is used to explain a deep attraction to and need for contact with nature. This is an instinctual and possibly inherited affinity for nature that goes back to our early evolution when nature was a resource, habitat, shelter, and place of awe for our ancestors. This need to experience the pleasure and phenomenon of nature lies within our DNA. Of course, when we don't have access to nature, or stars, or wildlife, we may be less fulfilled and feel their absence in our lives. This can be exacerbated by our environments, particularly urban environments that lack abundant, fecund, and accessible nature and/or the ability to engage with it.

Chronic stress is the antithesis of a peaceful state. Not only is it anxiety producing; it has several physiological consequences, including diminished immunity and antibodies, high blood pressure, artery-clogging deposits, and brain changes that may contribute to depression and substance abuse. Research also suggests that stress may directly cause people to eat more and indirectly decrease their sleep quality and quantity and exercise (Harvard Medical School, 2020). Testing for stress is relatively easy, as elevated cortisol levels indicate higher stress and are easily measured through saliva tests. To counteract stress, physicians have suggested using relaxation responses, including focusing on certain words such as peace or calm or visualizing tranquil scenes (Harvard Medical School, 2020). Another beneficial response is to spend time in nature, a garden, meadow, forest, or wetland, to immerse the body and mind in an intensive biophilic experience. There is a preponderance of data documenting the stress-reducing abilities of nature through both physical and visual interactions, resulting in decreased blood pressure, lower cortisol, reduced sweat production (sweat is a biomarker for stress), anxiety, and more. To say that a nature experience can reduce stress and enable a more peaceful and healthier state is clearly substantiated, and those seeking a state of peace often venture into "big" nature. Many of us choose to spend our precious vacations in emblematic immersions on the Pacific Crest Trail, commit to gardening as a passionate avocation, or spend a significant amount of our time in urban parks and greenways.

If pain equates with a non-peaceful state, exposure to nature can be the balm. Studies have shown that when a patient recovering from gallbladder surgery has a view of nature instead of a brick wall, the nature-viewing patient required less pain medication, had shorter hospital stays, and was less challenging and had a better attitude based on the care providers' observations (Ulrich, 1984).

Other evidence from the work of Kuo and Sullivan documents that having trees in a residential environment, in this case public housing projects in the city of Chicago, resulted in less aggressive behaviors, a reduction in crime, a deeper sense of belonging, and pride of place. The only demonstrable difference in the settings was the presence of trees (Kuo & Sullivan, 2001). As noted earlier, one aspect of peace is an absence of aggression and violence. Here in a housing project, sadly often stereotyped as a place of stress and crime, the reduction of these conditions was achieved with a relatively modest cost by planting and preserving trees. There are, of course, many other benefits, including the shade produced by the canopies, and we know that aggression increases in hotter environs. The increased habitat and wildlife; a softening of the aesthetics; increased fascination with the environment; and a diversity of engagements for the children such as collecting twigs and leaves, playing games amongst the trees, and climbing them all decrease stress.

Restoration attention theory (ART) was developed by Kaplan and Kaplan (1989). The ART hypothesis is that exposure to nature has the capacity to restore one's ability to focus. The premise is that everyday tasks, particularly work tasks, but also other stressors as well, decrease our ability for continuing attention, and, consequently, we lose efficiency and competence and are at greater risk for making mistakes. The consequences can in some situations can be catastrophic, such as in medical facilities. ART also has many applications for people experiencing trauma, anxiety, PTSD, and other emotional or psychological stress, as ART strengthens our ability to function in a more connected and focused manner, developing the strength to cope and face difficult challenges.

The use of gardens to elicit a sense of peace is expanding from what were originally interpreted as "healing" gardens to gardens with more specificity, for example, those that attempt to address trauma-informed design (TID). Some of the early models were gardens designed for veterans suffering from symptoms caused by combat trauma. As described by Winterbottom and Wagenfeld (2015):

> Post-traumatic stress disorder is triggered by exposure to a terrifying experience (stressor): responses to that stressor range from helplessness and hyperactivity to intense fear or horror. Symptoms associated with PTSD are intrusive, recollection, avoidant/numbing, and hyperarousal. Sensory gardens for veterans with PTSD, usually located at VA hospitals and clinics, military installations, or memorial sites, must be defensible spaces: for veterans or anyone with PTSD, the fear of recurring trauma is real.
>
> *(p. 183)*

The use of gardens to attain and celebrate a spiritual peace is found across the long history of landscape architecture, in built environments, churches, and monasteries and in outdoor environments. Examples are the Buddhist tradition of marking sacred trees in the forest or Native American cultures that designate sacred natural places that only tribal members can visit for prayer and ritual. Most monasteries had gardens where medicinal herbs were raised, as monasteries were essentially the first hospitals prior to any organized healthcare delivery. The monastery garden was also a place of cultivation for fruit trees and vegetables and functioned as a symbolic space, where the green lawn

FIGURE 3.2 In stress-inducing environments, such as Veterans Affairs (VA) hospitals, gardens can meet the needs of diverse user groups, such as offering space for contemplation for patients, workspace for care providers, and meeting and social places for staff.

represented rebirth and an everlasting life. In the early monasteries, the garden was twofold, a place of work and cultivation and a space to practice the monastic virtues humility and honesty. Monks could find peace within the garden and practiced attention restoration prior to returning to their studies and chores. The plan of Saint Gall, around 820 AD, represents an idealized plan of a monastery garden with many garden typologies, including a cloister garden, medicinal herb garden, vegetable garden, and a fruit garden that also served as a cemetery. The gardens are most interesting for the mixed aesthetic and functional plantings; for example, the herb garden contained ornamental plants mixed with herbs and spices. In addition to monastery gardens, other cultures have used the garden as a place of spiritual fulfillment and as a symbol of paradise – a place of peace and respite from the stresses of daily life and the earthly world.

The Islamic-style garden, displayed today in the Court of the Lions and the Court of the Myrtles in the Alhambra Palace and the Court of the Water Channel at the Generalife, date back to the 14th century in Granada, Spain, and are found throughout North Africa, the Middle East, Asia, and India. The Islamic garden, unlike the European garden that was designed for walking and exuberant floral displays, was instead intentionally designed for rest, reflection, and contemplation. The garden is organized around the Charbagh, a quadrilateral layout with four primary spaces divided by walkways or water runnels. There are many stylistic varieties, Umayyad, Abbasid, Mughal, Ottoman, and others, but all represent paradise, dating back to the reference in the Qur'an that believers will dwell in gardens which have rivers flowing beneath them (Aga Khan Centre, 2023). The garden, a metaphorical paradise, was also a tangible manifestation as the verdant gardens replete with flowing waters, shade, and fruit trees were a respite from the hot, dusty climates where many of the Islamic gardens are found. The garden is surrounded by a wall, ensuring it as a place of safety, and offers a peaceful, sensory experience. The water reflects the heavens and trees and provides a calming sound to stimulate inner peace. Many of the plantings are aromatic, and the scent amplifies the sensory experience, as exemplified by the lemons and oranges, herbs, roses, jasmine,

FIGURE 3.3 Islamic gardens are sensory enhancers that stimulate feelings of restoration, fascination, comfort, and a joy of life. The use of water and vegetation creates a "paradise" in contrast to the hot, dry context in which most are found.

Source: Spain (Granada) Alhambra-Beauty of Islamic calligraphy and arabesques around the arches (www.flickr.com/photos/126397594@N03/28406274320) by Güldem Üstün (www.flickr.com/photos/guldem/) is licensed under CC BY 2.0 (https://creativecommons.org/licenses/by/2.0/).

and lilies that are commonly found in these gardens. The Aga Khan Centre in London, England, had taken the traditional style of an Islamic garden, reinterpreted it as a 21st-century creation, and adapted it to the English climate. The peace and containment found within its six gardens is an apt counterpoint and de-stressor to the fast-paced urbanity surrounding it.

Landscape as a Supportive Space

Landscape fosters a sense of peace, as noted by the increasing amount of summer vacationers flocking to national parks to seek a transformational peaceful experience amongst wild nature – a peace that is both pure and spiritual and dependent on a nature-based experience, such as those found in ancient forests, in the blooming of alpine meadows, or deep within sandstone canyons.

Peace is also often an explicit intention of many landscape architects and designers. One notable example is the "peace park." Located between the borders of the Canadian province of British Columbia, Canada, and Washington state, USA, lies Peace Arch Historical State Park. The 22-acre strip of land abuts the Puget Sound was set aside as a day use park. Dedicated in 1921 to commemorate the Treaty of Ghent and the Rush-Bagot agreement that ended the war of 1812, it provided a peaceful resolution between the US and Britain and the unguarded Canadian border. A reinforced concrete Peace Arch monument, its largest feature, with one foot in each of the countries, sits on the border at the 49th parallel and is surrounded by extensive lawn defined by sweeping drifts of planting beds and large trees. This gardenesque treatment offers a diversity of spaces from large open lawns that host community celebrations and recreational activities to smaller niches within the garden beds that offer a level of privacy and intimacy. The space was originally intended as a rather formal commemorative landscape feature of the end of the War of 1812. It is now being adapted by families torn apart by emigration restrictions and being used as a place of congregation, a neutral space where family members in Canada can spend time together with those in the US despite that fact that they can't congregate in either country. During COVID-19, Peace Arch Park became the most frequent location for "border weddings" where both parties would meet at the same location without separation. Clearly this concept could be adapted to many places where divisions fester and families are burdened with anxiety, uncertainty, and often fear. Divisions in Cypress, between Mexico and the US, Ukraine and Russia, and Israel and Palestine are some of the places where such a park could bring a momentary peace for those disrupted by conflict or governmental policies.

Arguably better known, but in some respects not as unusual as Peace Arch Provincial Park in their use of land to create an open neutral territory between two countries, are two memorial peace parks in Japan. Nagasaki Peace Park and Peace Memorial Park in Hiroshima commemorate the horrific consequences that the dropping of two atomic bombs during World War II within these cities inflicted. Each contains architectural remains that convey the immense damage as well as elements of nature representing rebirth and peace. The parks are settings for many individual memorials to those who perished in the bombings, and both contain museums, water features, and green spaces. Each is a symbol of a national tragedy and preserved as a memorial dedicated to promoting peace and a future devoid of nuclear bombs. The Peace Flame at the Peace Memorial Park will stay lit until all nuclear bombs are decommissioned, and each year on August 6, the anniversary of the bombing, white doves are released during a memorial service.

Nature has long played a role in bringing literal and psychological peace to those suffering the insidious impacts of combat. Following the Civil War, many veterans were placed in hospitals that were intentionally sited in forested places to receive care. Letters written by the residents illuminate the sense of peace that these nature-based environments provided in their recoveries. Kenneth Helphand has extensively documented the purpose and power of gardens created within landscapes of war. These gardens offer both an escape to a better place and often elicit memories of home and a place of living versus death for those engaged in combat and struggling for survival. His documentation revealed that the gardens were created in the trenches of the First World War, and more recently created in Afghanistan and Iraq by American soldiers (Helphand, 2008). Following the publication of the book, he received an outpouring of communication from living veterans to share their individual stories of how a small modest garden meant so much and brought them, momentarily, a sense of peace and sanity. For those returning in body but struggling to readapt or coping with the loss of limbs and physical trauma, healing gardens play a critical role in their recovery. They are being introduced into VA hospitals and parks to provide a nurturing space to foster coping mechanisms and achieve a level of peace despite the veteran's shattering experiences.

Similar but generally smaller than peace parks is a form of spaces called "peace gardens" that have been emerging in response to street violence and loss of life perpetrated by law enforcement. They take the form of street gardens and are also termed "grief gardens," as they memorialize those killed by violence. Like memorials situated along roadsides marking sites of traffic fatalities, "peace gardens" tend to be vernacular in process and expression and are typically shrines that mark a spot upon which tragedy and violence forever changed the meaning of this space and community. These gardens offer comfort and memorialization and are often acts of protest against the relentless loss of life. They maintain some similarities with gardens that emerged in American urban neighborhoods in the 1970s and 80s that had been devastated by white flight, economic downturn, arson from landlords cashing in on insurance policies, increasing violence, a drug epidemic, and the aftermath of the urban riots of the 1960s. These spaces, abandoned lots, surrounded by war-torn landscapes, were repurposed as gardens. Many were named for their aspirational ideals such as Garden of Hope, El Jardine del Paraiso, and Tranquilidad. Ad hoc community gardens, they are typically created from salvaged materials, verdant found objects, and assemblages that express the creator's ethnicity and/or eccentricity. Most were transitory and unsanctioned, as are many of the "peace gardens." These gardens, such as "casitas" and other community gardens, offered rare access to a lush, green oasis, islands of calm and peace within very violent neighborhoods. Kenneth Helphand (2008) writes that gardens are effective due in part to the unique green materiality that comforts:

> Gardens can also be a form of consecration. An aspect of hope is that of solace, the giving of comfort, especially to those who are grieving or wounded. . . . The mechanism may seem mysterious, but perhaps it is the sense of refuge, of calm and peacefulness, or just the quiet and the connection to nature – to life.
>
> *(p. 225)*

FIGURE 3.4 As many New York neighborhoods fell into ruin and crime levels rose, "casita" gardens emerged as safe zones for community members, offering a sense of sanctuary similar to the "paradise" concept found in Islamic gardens, albeit grittier. These are self-constructed, managed green spaces where community, culture, and peace are celebrated.

Gardens are increasingly being inserted into landscapes of incarceration. As we gain a better understanding how exposure to nature can reduce aggression and stress (see Kuo & Sullivan, 2001; Young, 2004), they seem to become more applicable in carceral environments where these characteristics are magnified and depression and anxiety are rampant. Winterbottom and others have found the term "peace" used frequently when incarcerated individuals describe the effects of a garden on their psychological and spiritual states. Some formerly incarcerated individuals have described the garden as a space that offers peace, a rare experience when compared to the constant noise levels and tension typically found in many carceral environments. For many, the carceral garden is considered a sacred space. Some have described the carceral garden space as a safe zone where the intent of peace and contemplation is respected, and gang aggressions and hostel behavior are left at the garden gate (Winterbottom, personal communication, 2022–2023). At the Oregon State Penitentiary in Salem, Oregon, a large Japanese garden was created by the incarcerated members of the Asian American club. The participating incarcerated individuals helped to design and build a place within the prison where racial, gang, and predatory behaviors are forbidden and peace is attainable. Other incarcerated individuals have described those gardens as sacred places, not unlike a church or library, where silence pervades the space and users speak in hushed tones. Some carceral gardens are used to support restorative justice activities where an offender works to understand their actions and transform themselves into a different person, learning skills to resolve their anger, trauma, or aggression. The intent is to enable them to reform their self-image, become more peaceful in their actions and relationships, and seek peace in their lives. In his autobiography *Long Walk to Freedom*, Nelson Mandela (2008) reflected that over his many years of imprisonment, a garden offered what little else could:

> A garden was one of the few things in prison that one could control. To plant a seed, watch it grow, to tend it and then harvest it, offered a simple but enduring satisfaction. The sense of being the custodian of this small patch of earth offered a small taste of freedom.
>
> *(p. 926)*

Nature interactions, time spent, and lessons learned in the garden are used to help released incarcerated individuals make the transition from inmate to freedom and develop healthy, trusting relationships through nature-based activities (Clayton, 1919).

In *Gardens of Defiance* (2008), Kenneth Helphand describes how modest garden plots were created in battle zones from the trenches of WWI to the more recent conflicts in Iraq and Afghanistan. These small gardens, carved from landscapes of devastation, were places that offered counterpoints to the

FIGURE 3.5 The healing garden at Oregon State Penitentiary is a safe space, built by incarcerated individuals and conceived by the Asian American club within the institution. It is both a metaphorical and applied healing space, full of symbolic references and used by mental health providers to calm patients experiencing psychotic events.

Source: Randy Guzek, Memorial Healing Garden Director.

FIGURE 3.6 Water and wildlife are highly desirable elements in healing gardens. In carceral settings, they convey hope, reconnection, and a sense of peace to those in sensory-deprived, highly stressful environments.

Source: Randy Guzek, Memorial Healing Garden Director.

horrors of the battlefield and reminded soldiers of their homes and families and something to nourish as opposed to kill. Clayton (1919) articulates the profound importance the garden played in the lives of those immersed in combat:

> Soldiers hoped to be spared from death, and they hoped for peace. The familiar cycle of life – birth, growth, maturity, and death – is perhaps nowhere more visible than in a garden, where the cycle plays itself out each season and over the years. But the cycle cannot be ignored in battle, where perhaps the strongest emotion is the soldiers' fear of dying. By immersing themselves briefly into the garden world, however small, soldiers-maintained contact with the most immediate part of their humanity: that of being alive. *(p. 72)*

Gardens are also important in the process of reconciliation and in achieving a peace between warring factions after a ceasefire is put in place or the conflict has ended. Shortly after the war in the Balkans came to a fragile end in 1994/5, many communities and families were split apart following the ethnic lines that had been drawn across the former Yugoslavia, resulting in the genocide of Bosnians, Serbs, and Croatians. The Community Gardening Association in Bosnia and Herzegovina (CCABH) was founded and directed by Sarajevo native and war survivor Davorin Brdanovic to bring the formerly warring ethnicities together in a series of gardens to share in the fruits of labor, engage in social interactions, and find peace for those severely scarred from the ravages of the war. Davorin attempted to recreate the prewar ethnic makeup of the local community wherever possible among the garden participants. This resocialization and reconciliation effort was supported by the CCABH, who provided material assistance in land clearing, including live mines, and permission to use the land, as well as providing seeds, soil amendments, and tools to grow food, which was very scarce after the bombing abated and because many had lost their homes. These multiethnic gardens produced not just fruit and vegetables but also therapeutic spaces to accept former opposition members

FIGURE 3.7 In this postwar garden, former Bosnian, Serbian, and Croatian enemies share coffee, time, and gardening activities together as they reconstruct their lives and reconceptualize their neighbors not as threats but as fellow humans who shared in their suffering.

back into one's life through gardening; they could come to understand that these were human beings, not an autonomous fighter. Davorin noted,

> one man came to the garden and couldn't speak, was essentially non-verbal for a year. Through the gardening and his developing a trust over time with those of the same ethnicity as the perpetrators, he was able to overcome his paralysis and speak again, becoming a regular who gardened every day and over the years developed deep friendships with many of the other participants.
>
> *(Winterbottom, personal communication, 1998–99)*

Memorialization

For many, memorials represent places where peace and the restoration of one's mental and emotional health and well-being can be accessed. They provide safe zones to grieve and confront personal obstacles and find a pathway to a more peaceful existence. Often, they incorporate nature as both a therapeutic mechanism and symbolic element that references concepts of peace, hope, and sanctity. Many serve as large parks in which one can escape urban stressors and offer a moment of escape to focus, meditate, center, and pay respect to those who have sacrificed or simply passed on. Part of the relatively new spatial concept of memorials, versus the object as memorial, can be traced back to Saint Gauden, a 19th-century sculptor, who created many memorials and moved from object to spatial expression, "Saint Gauden's innovations were part of a broader shift in the conception of the public monument, from an object of reverence to a space of subjective experience" (Savage, 2010, p. 12). Well-known examples include Arlington National Cemetery, a site where those who have served are laid to rest. The Mount Auburn Cemetery, in Cambridge, Massachusetts, is not specifically dedicated to veterans, although many are buried within its grounds, but is one of the first rural garden cemeteries created in the United States. At Mount Auburn, beauty and serenity are values, and, as their mission states, their focus is on "creating and celebrating beauty and fostering a place of serenity for all who seek it. Evoking the essential experience of Mount Auburn as a landscape, a place of beauty, and a source of serenity. Nurturing our defining aesthetic – human experience and nature together" (Mount Auburn Cemetery, 2020, Mission & Vision). As a publicly accessible garden, access to a peaceful state is offered to all and as such is more holistic in its purpose. Many come to pay respects to those laid to rest within its grounds, while others simply stroll through the space seeking the calm and restful state within its botanical collection that includes lakes, ponds, significant and massive trees, seasonal plantings, swaths of lawn, and undulating topography. Both Arlington and Mount Auburn exemplify the role that burial grounds offer across cultures and a wide spectrum of users. They offer connection to the cycle of life and, of course, death, a space to connect to the spiritual mysteries and the cycles that align with certain belief systems. Many burial grounds incorporate the idea of peace and serenity more specifically in their naming protocols, such as Home of Peace Cemetery in Lakewood, Washington, and Los Angeles, California, or Mount Peace cemetery in St. Cloud, Florida. The importance of these naming conventions and the stated missions of these burial grounds is that the environments are intentionally treated in a gardenesque manner and intended to offer a sense of peace as a counterpoint to a tragic event, such as the passing of a friend or family member.

Other forms of memorialization provide spaces to grieve and reflect. Many of these are responses to catastrophic events that inflict great loss, societal

FIGURE 3.8 As a sacred space of both departure and visits, the garden cemetery model uses an abundance of plantings to create a peaceful setting to grieve and pay homage to those who have passed. The natural, therapeutic environment supports the sacredness and purpose of the place.

schisms, and disruption. These memorials are intended to ensure that those lost will be remembered and/or honored and that those who are grieving can find peace and solace. Arguably the best-known American example is the Vietnam Memorial, designed by Maya Lin, who was an architecture student at the time at Yale University and created a memorial quite strikingly different from any that predated it. In her statement of submittal for the design competition, quoted in Scully (1991), she noted,

> It brought to a sharp awareness of such a loss, it is up to each individual to resolve or come to terms with this loss. For death is in the end a personal and private matter and the area contained within this memorial is a quiet place, meant for personal reflection and private reckoning.
>
> *(p. 28)*

The memorial is minimal in its form and representation. It has been described as a slice through the earth. In essence, it is a retaining wall clad in black polished granite wall that bears the names of all of those killed during the Vietnam war. A sloping path guides the visitor downward as they enter the space and ascend out as they depart. The entire procession, including the viewing of the names, offers a catharsis for the pain of loss for those returning from the war and for family members and friends of those who paid the ultimate price. The space is somber and processional, and many visitors take rubbings of the names home and/or leave notes and offerings below the name. Others sit upon the grass within the triangular space, to mourn, grieve, and spend time with the memory of those who have departed. The space is defined on two sides by the wall and by groves of trees enclosing the upper space that provide a verdant backdrop to the memorial and a therapeutic green space, providing peace from what for many was an incomprehensible loss of life with little justification or strategic purpose. Many visitors have reflected upon the sense of peace they were able to find there. For example, one visitor noted on the Vietnam Veterans Memorial Fund (VVMF) blog (2022):

> I have visited The Wall twice, many years ago. It of course brought tears for all those we lost to that awful war, which I was able to watch on the evening news, to my horror. I don't think any of us who lived through it have ever been the same. It also brought a sense of peace to me that these people were getting some recognition for their sacrifice.
>
> *(quotation #19)*

To find that sense of peace, and to be able to release pain and sorrow, a place was needed, a place that was sacred ground, and the etching of 58,000 names into a black wall bestowed a power upon that site. Upon the wall, they have come together, brothers and sisters, fallen but not forgotten, and this place calls to those in need, and they come to shed their pain and find acceptance, peace, and validation of their pain. Will this function change as time goes on after the immediate survivors have passed and time blurs their connections, still so raw to a mother, precious to a child, and bittersweet to a partner? While the relationships may change, the sense of peace and connection to the existential meanings of life and death will endure, and this will continue to be a site of peace, sorrow, and meaning.

Sacred groves have long had meanings associated with them. For the Greeks, they were dedicated to a god or gods. In India, the ***kovil kaadugal*** (temple forest) held religious significance for Hindus, and the Celts performed rituals based on their mythology in their sacred groves. The sacred grove is referenced and/or found in Asia, Africa, Europe, and the United States. Some memorials are intentionally placed within these groves. The AIDS memorial, in San Francisco, California, is aptly named The Grove, and this 10-acre memorial is the United States' only federally designated national AIDS memorial. As described on the National AIDS Memorial (n.d.) website, this memorial is:

> a dedicated space in the national public landscape where anyone who has been touched by AIDS can grieve openly without being stigmatized, can find comfort among others whose lives have been affected by AIDS, and can experience the feelings of renewal and hope inherent in nature.
>
> *(para. 1)*

The memorial consists of a paved gathering circle inscribed with the names of those who passed that contains various boulders with phrases and poems carved into their surface. The memorial is nestled into an actual grove of redwood trees and ferns in San Francisco's Golden Gate Park. The National AIDS Memorial brings those who have been stigmatized and devastated by the plague, initially ignored and overlooked, together to share their bonds, grieve for those who were taken, and find solace and pride in their culture and a sense of healing from the traumatic history of the disease. Being amongst the trees is an essential therapeutic benefit found within the memorial. A small stream generates a soothing sound; the perennials and ferns lend a softness to the environment and offer habitat that draws wildlife drawn to this area. The towering redwoods elicit awe and a connectivity to the primal world of nature and a spiritual presence that can soothe those in pain or buried in sorrow.

The previous examples are memorials that help users attain a sense of peace. Most have a focus on an event, war, a plague, or a burial site. The Memorial for Peace and Justice is an intentional memorial to peace, or so the name would suggest, but on their web site under mission and goals, the word peace never appears. The following is one excerpt from their mission: "The National Memorial for Peace and Justice provides a sacred space for truth-telling and reflection about racial terrorism and its legacy" (The Legacy Sites, 2023).

FIGURE 3.9 Within a grove of trees sits the AIDS memorial. Nestled in Golden Gate Park, the memorial is a place to honor and remember those who were lost to the illness. While somber, it celebrates the communities devastated by HIV, AIDS, and politics that surrounded the crisis. This natural setting conveys a sense of peace to those who survived the disease and those who were irreparably changed by it.

Source: "National AIDS Memorial Grove in San Francisco's Golden Gate Park in March 2018" (https://commons.wikimedia.org/w/index.php?curid=67301323) by Gregory Varnum (https://commons.wikimedia.org/wiki/User:Varnent) is licensed under CC BY-SA 4.0 (https://creativecommons.org/licenses/by-sa/4.0/?ref=openverse).

Clearly a primary intention is to awaken visitors to a truth long ignored and unspoken and to reconcile the divisive history of race in America and forge a peace between the races. The memorial, conceived by Bryan Stevenson, a public interest lawyer and founder and executive director of the Equal Justice Initiative, a human rights organization based in Montgomery, Alabama, is an extraordinary project that confronts the history of lynching and racial injustice and provides a space where the story can be told. Communities where a documented lynching has taken place can bring a duplicate steel marker back to their communities and install it at the lynching site to acknowledge, heal, and promote peace amongst the racial divisions that remain. This memorial brings to the fore the complexity of creating places intended to forge a peace between and within individuals holding diverse perspectives. Is peace the goal, as the title would suggest? Is it reconciliation, justice, remuneration, or acknowledgement? Does peace in this context mean equal rights, resolution of conflict, or forgiveness? Given the horror and trauma inflicted through lynching, what does/can peace look like? Does peace come from the perpetrator, the victim, or, as restorative justice suggests, a reconciliation between both parties, where acknowledgement, forgiveness, redemption, and transformation become the pathways to peace? Peace in these and many other contexts is complicated, essential, and difficult.

The term peace garden has also been used to create more modest but equally important spaces of healing and commemoration. The city of Minneapolis, Minnesota, features several notable examples that appeared following the murder of George Floyd in 2020. These spaces, created by community members to honor victims of violence, reveal the connection between acts of violence to larger issues of systemic racism, inappropriate use of violence by law enforcement, and the inequities faced by many in neighborhoods historically underserved and suffering greater rates of violence and racial discrimination. The Philando Castile Peace Garden was created by members of the Falcon Heights community, local organizations, and members of Philando Castile's family as a "contemplative gathering space for peace, justice, and healing" (Seeds of Peace Philando Castile Peace Garden, 2024, Who we Are). One enters a winding pathway, beginning where the policeman initially pulled the car over and continues weaving between trees and sloping up until it terminates at a gathering space. The circular gathering area is etched with a cosmogram oriented towards where Philando's car had been parked when he was killed. Carved into the paving are the words "birth" and "death," and a line joins the two after encircling a tree that symbolizes rebirth. Five stairsteps connect the plaza to a sidewalk below, representing the five shots that killed Mr. Castile, and a curved bench defines the outer edge of the plaza, with its vertical back slightly tilted at the same angle as was Philando's seat when he was killed. Local politicians did not want Castile's name permanently etched into the memorial, nor any references to the cause of his death, yet clearly those in the community and visitors understand what had happened and come to visit the site to pay homage to Philando Castile. Flanking the winding pathway are West African adinkra symbols made from steel that represent peace and tranquility. As described on the Philando Castile Peace Garden website (n.d.), that oversees the garden today this garden incorporates both symbology and continuing community participation:

> The Community Peace Garden has been designed to feature and preserve (adinkra symbols) art pieces (and memorial stones) created and placed there by community members in the wake of Philando Castile's tragic death, and to allow families and community members, law enforcement, and local, state, national and international leaders to join together in the search for peace and justice.
>
> *(para. 3)*

Many who come leave objects of remembrance, including painted stones, flowers, and keepsakes. On the Saint Paul and Minnesota Foundation's (2024) webpage on the garden, Philando's mother Valerie Castile reflected:

> We wanted this place to become a place where you can relax, feel safe, contemplate on yourself, do some self-inventory. A place to be vulnerable, a place to transform yourself, where there are so many characteristics and attributes, they're here.
>
> *(A Mother's Love)*

The Philando Castile Peace Garden is a powerful representation of the community's need for healing spaces located on the sacred ground of shared trauma. If not for these efforts, these sites would remain cleaned up and unnoticed, leaving the community without a space to mourn, heal, and find peace. As Sharon Castle (Saint Paul and Minnesota Foundation, 2024), one of the early garden supporters, noted,

> Falcon Heights, this community where I've lived for 27 years, will forever belong to the long list of places where black people have been shot and

FIGURE 3.10 The Philando Castile Peace Garden, Minneapolis, Minnesota, honors the life and death of Philando, killed by a police officer in 2016.

killed. As we move forward, we can't forget what happened here. The Peace Garden provides a calm place along a very busy street to stop and reflect. When I visit, I often have conversations that I wouldn't have had otherwise. It's a gift from Philando and Valerie to have this space right here.

(The Inspired Community Member)

In response to the murder of George Floyd, other spaces in Minneapolis, the George Perry Floyd Square and the nearby Say Their Names Memorial, were designated as sacred spaces of peace and healing. George Perry Floyd Square lies at the intersection where George Floyd was killed. A block anchored at each end with large artworks depicting raised fists at two intersections is marked with the names of people of color who have been killed. These names are repainted each year and additional names added. A raised garden, wall art, murals, and a shrine at the location where George Floyd was killed containing many tributes and offerings are some of the many interventions created along the one-block square. The space is a place of both healing and peace and commemoration and resistance. Many of the abutting storefronts are now galleries that display images related to the Floyd murder; others are places

FIGURE 3.11 A temporary art installation, the "Say Their Names" cemetery, memorializes the story of black lives lost to violence inflicted by law enforcement in America. The creators are trying to make the installation permanent on the site.

of community art production. An abandoned gas station renamed "Peoples Way" is a gathering space with RIP murals painted on the building where vigils and conversations are held. The square is very much a collection of memories honoring those who have died as well as a political statement expressing the community anger at the inequities and injustices, a place of exchange and storytelling, and a place of healing as more recent passings continue to be honored. The Say Their Names memorial, created by Anna Barber and Connor White, is a symbolic neighborhood cemetery containing 100 headstones with the names of people of color who lost their lives at the hands of the police. It is nestled into a low-lying lot, with trees flanking the perimeter, and is a unique place of solitude and contemplation. The naturalistic setting harks back to the relationship of burial and landscape and the importance of nature as a source of peace and healing. A small water body runs along one side of the site, and, in comparison to George Perry Floyd Square that reflects the urban street typology, the Say Their Names memorial is notably more quiet and softer, a supportive natural environment that offers calming and sanctuary as a counterpoint to the actual content, which is overwhelming and traumatic. Both memorials are cries of outrage that bring those who visit these spaces both peace that their legacies are being honored and indignity that these acts are still being perpetrated upon people of color.

Design Strategies

Are there elements, spaces, forms, and materials that convey, reference, or elicit qualities of peace, symbolically, emotionally, spiritually, and psychologically? Are some more effective than others, how do cultural preferences affect these choices, and are there universal meanings associated with these choices? These topics are important, and unfortunately there is surprisingly little research completed around these or similar questions, and thus evidence-based responses are rarely possible. We do, however, have some evidence that forms and materials can elicit calming effects, reduce stress and anxiety, and enable one to reach a more peaceful state within certain environments. For example, researchers report that serpentine forms, pathways, planting masses, and structures such as pergolas are more effective at creating a more peaceful and calming state (e.g. Ibarra et al., 2017). The flow of these forms and their being less angular and more meandering seem to have a beneficial, calming, therapeutic effect. In rankings of materials that offer therapeutic benefits in the landscape, water and plants have ranked as the most desirable in several studies. The exclusion of hard, manufactured surfaces such as concrete and the use of natural materials, stone, wood, and earth forms are generally preferred, as are aromatic, textural, and seasonal plantings. Wildlife, such as butterflies, birds, and small mammals, are often mentioned in participatory design sessions, and the natural sounds of birds, moving water, and plant and leaf rustlings are often desired, especially in urban settings. The use of pollinator plants that offer food, such as bell-shaped flowers for hummingbirds and berry plants for songbirds and other species, will increase these populations within the garden. Sensory gardens are often described as calming and transformative and, as a distraction, offer an escape from stress, PTSD, and trauma. Veterans often talk about the harmony that dense vegetation with textural and color diversity can elicit, and they provide an escape and distraction from the memories and flashbacks of war. Most prefer gardens in which they can physically and psychologically immerse themselves, often preferring nature in lieu of social interactions, art, or architectural features. That noted, some have mentioned specific plants that are prevalent in the battle zones they occupied and had negative connotations for them. Bamboo and banana plants, for example, are a few that Vietnam veterans have asked that we not use in their healing gardens.

Color has often played a role in "peace" gardens, with white being frequently used, as it symbolically references peace, purity, and innocence. White or pink Asiatic lilies, many of which are strongly aromatic, are also a symbol of peace for many cultures and are used in peace gardens. As mentioned, water is a material frequently used in memorials and therapeutic gardens. This is often in the form of large reflecting pools, waterfalls and spillways, and water walls. Water, as the life-giving force, has spiritual and emotional properties, reflects light, becomes animated and mesmerizing, creates a range of soothing sounds, attracts wildlife, and supports plants that many cultures desire such as lotuses, water lilies, taro, cypress, and willow trees.

Conclusion

The pursuit of peace as an intentional outcome in landscape architecture is quite common. The concepts of biophilia, attention restoration, and increased well-being from spending time in nature-based places have an obvious reliance and influence on those entering the profession. Peace may not be the explicit primary goal in landscape architecture projects, for example, urban development projects, work environments, commercial projects, and others, but certainly the conscious effort to use nature to diminish the negative effects of urban noise, air pollution, gray, and concrete is becoming an important goal in landscape design. The intentionality of bringing peace as a concept to the fore is arguably more evident in memorials and "peace" gardens, as discussed. The need for peace and aspirations to be explicit about its importance are also prevalent in other forms of garden typology. Vernacular projects, which are created not by landscape architects but by individuals and communities, are designed by those who perceive a need for peaceful places. Declarations of peace versus violence use gardens as a form of resistance to injustice. Carceral and casita gardens, "peace" parks and "war" gardens created in battle fronts and refugee camps, are safe "peace" zones of resistance to violence and injustice and as a sanctuary from the surrounding context. Many attempt to impart an alternative and/or moral positioning against the pervasiveness of violence and conflict within the community or, in the case of soldiers, an escape to a saner place in opposition to what they are asked to do and to survive. Other gardens and landscape spaces intended to convey a peacefulness, a place of contemplation and escape, and a place to connect to a deeper, possibly spiritual realm are provided in sacred church gardens, cemeteries, and memorial landscapes.

References

Aga Khan Centre. (2023). *Islamic gardens*. https://www.agakhancentre.org.uk/islamic-gardens/

Beckman, C. (2014, July 27). Let peace begin with me. *Catholic 365*. https://www.catholic365.com/article/38/let-peace-begin-with-me.html

Clayton, P. B. (1919). *Tales of Talbot house: Everyman's club in Poperinghe & Ypres, 1915–1918*. Chatto & Windus.

Harvard Medical School. (2020, July 6). Understanding the stress response. *Harvard Health Publishing*. https://www.health.harvard.edu/staying-healthy/understanding-the-stress-response

Helphand, K. (2008). *Defiant gardens*. Trinity Press.

Ibarra, F. F., Kardan, O., Hunter, M. C. R., Kotabe, H. P., Meyer, F. A., & Berman, M. G. (2017). Image feature types and their predictions of aesthetic preference and naturalness. *Frontiers in Psychology, 8*(632). http://doi.org/10.3389/fpsyg.2017.00632

Kaplan, R., & Kaplan, S. (1989). *The experience of nature: A psychological perspective*. Cambridge University Press.

Kellert, S., & Wilson, E. O. (1995). *The biophilia hypothesis*. Island Press.

Kuo, F. E., & Sullivan, W. C. (2001). Aggression and violence in the inner city: Effects of the environment via mental health. *Environment and Behavior, 33*(4), 543–547. https://doi.org/10.1177/00139160121973124

Mandela, N. (2008). *Long walk to freedom: The autobiography of Nelson Mandela*. Hachette UK.

Merriam-Webster. (2023). Peace. In *Merriam-Webster.com dictionary*. https://www.merriam-webster.com/dictionary/peace#:~:text=Legal%20Definition-,peace,for%20by%20law%20or%20custom.

Mount Auburn Cemetery. (2020, October 1). *Mission & vision*. https://www.mountauburn.org/mission/

National AIDS Memorial. (n.d.). *The grove story*. https://www.aidsmemorial.org/grove-history

Olmsted, F. L. (1997). Public parks and the enlargement of towns. In S. B. Sutton (Ed.), *Civilizing American cities: Writings on city landscapes* (pp. 52–100). Da Capo Press.

Philando Castile Peace Garden. (n.d.). *Seeds of peace*. https://philandocastile-peacegarden.org/

Saint Paul & Minnesota Foundation. (2024). *Turning tragedy into healing*. https://www.spmcf.org/blog/philando-castile-peace-garden

Savage, K. (2010). The obsolescence of sculpture. *American Art, 24*(1), 9–14. https://doi.org/10.1086/652736

Scully, V. (1991, July 14). The terrible art of designing a war memorial. *New York Times*. https://timesmachine.nytimes.com/timesmachine/1991/07/14/issue.html

Seeds of Peace Philando Castile Peace Garden. (2024.) *An invitation*. https://philandocastile-peacegarden.org/

The Legacy Sites. (2023). *The national memorial for peace and justice*. https://legacysites.eji.org/about/memorial/

Thirdblade Photography. (2014). Fall in Central Park [Photograph]. *Flickr*. https://www.flickr.com/photos/49115235@N06/15746400912

Ulrich, R. S. (1984). View through a window may influence recovery from surgery. *Science*, *224*(4647), 420–421. http://doi.org/10.1126/science.6143402

Varnum, G. (2018). National AIDS memorial Grove in San Francisco's Golden Gate Park in March 2018 [Photograph]. *Wikimedia Commons*. https://commons.wikimedia.org/w/index.php?curid=67301323

Vietnam Veterans Memorial Fund (VVMF) (2022, August 31). 40 quotes about the Vietnam war. *Your Stories Your Wall*. https://vvmf.wordpress.com/2022/08/31/40-quotes-about-the-vietnam-veterans-memorial/

Winterbottom, D., & Wagenfeld, A. (2015). *Therapeutic gardens: Design for healing spaces*. Timber Press.

Young, C. (2004, September 1). *Green peace*. https://greatergood.berkeley.edu/article/item/green_peace

4

DESIGNING FOR PEACE THROUGH A TRAUMA-INFORMED PERSPECTIVE

Erin K. Peavey and Rebecca L. Ames

In today's complex and fragmented world, the need for peace has become increasingly vital. While we often think of peace in the context of global conflict and diplomatic relations, the true essence of peace extends far beyond these parameters. Peace is also a deeply personal experience. And when it is disrupted, it can create ripple effects on our personal well-being and on our ability to form meaningful relationships with others and with the world around us.

Those of us who have experienced trauma know that these experiences can dramatically change the course of our lives and the way we see the world. Over the last four decades, the study of stress and trauma has grown into a transdisciplinary field incorporating research across neuroscience, biology, social science, healthcare, and public policy. As such, the word 'trauma' has acquired many definitions. For the purposes of this chapter, we will refer to the definition supplied by the Substance Abuse and Mental Health Services Administration [SAMHSA], 2014):

> Trauma results from an event, series of events, or set of circumstances that is *experienced* by an individual as physically or emotionally harmful or life-threatening and that has lasting adverse effects on the individual's functioning and mental, physical, social, emotional, or spiritual well-being.
> *(p. 7)*

The adverse or negative impacts of traumatic experiences are often attributed not to the event itself but rather to how the event is perceived or experienced by the individual (SAMHSA, 2014). For example, did a child interpret their father's absence as a lack of safety and love, or as a temporary condition

DOI: 10.4324/9781003462019-4

unrelated to their worth? This process of meaning-making depends on several factors, such as the availability of social support, the developmental stage of an individual at the time of the traumatic experience, and, importantly, the physical environments an individual interacts with.

While once believed to impact just a small subset of the population, it is now understood that the experience of potentially traumatizing events, particularly in childhood, is relatively common. In the United States alone, approximately 64% of adults report having experienced at least one type of adverse childhood experience (ACE), such as abuse, neglect, or witnessing domestic violence, before the age of 18, and nearly one in every six adults report experiencing four or more types of ACEs (Centers for Disease Control and Prevention [CDC], 2023). These early experiences have been linked to psychological and physiological challenges that can persist over a lifetime, influencing our state of internal peace and our efforts to create a more peaceful world. Recent events – from the pandemic and rising social isolation to experiences related to the climate crisis – have contributed to an unprecedented jump in mental health concerns in recent years. Globally, more than one in eight people are living with a mental health disorder – such as anxiety, depression, bipolar disorder, and post-traumatic stress disorder (PTSD) (World Health Organization [WHO], 2022).

While the precise influence that the built environment can have on our mental and emotional health, as well as our internal and external peace, is still being understood in the context of broader public health initiatives (Núñez-González et al., 2020), we know that the environment matters (Evans, 2003). From the layout of a room to the planning of neighborhoods and cities, design has the power to enhance or impede our internal sense of peace and our connection with others and the world around us. Designing for peace recognizes that human flourishing goes beyond designing for survival and material needs, emphasizing the importance of creating environments that support individuals' holistic well-being, enabling them to thrive emotionally, mentally, and socially.

This chapter will begin to answer the question: How might the design of the built environments that form the backdrop of our lives be created to reduce the burden of stress, trauma, and mental illness and promote peace and well-being among individuals and communities around the world? We will examine this question through the lens of trauma-informed design (TID), an emerging field that builds on research from environmental psychology and evidence-based design to account for the specific needs of individuals who may have experienced trauma. TID offers a framework through which we can begin to consider how the design of the built environments that surround us can help to restore a sense of safety, promote healing, and empower individuals and communities.

Your Brain on Trauma: a.k.a. What's Neuroscience Got to Do With It?

As Dr. Gabor Maté, renowned addiction, childhood trauma, and mind-body health expert, stated in an interview:

> Trauma is a psychic wound that hardens you psychologically that then interferes with your ability to grow and develop. It pains you and now you're acting out of pain. It induces fear and now you're acting out of fear. Trauma is not what happens to you, it's what happens inside you as a result of what happened to you.
>
> *(HumanWindow, 2019, p. 4:10)*

Before we can understand the ways in which the environment can be designed for peace, we must first understand how our peace is disrupted, how the things we experience throughout our lives can burrow their way into our psyches and fundamentally shift our relationship with the environment around us. First, although the experience of harmful or potentially traumatic events in childhood is nearly universal, human beings are resilient, and these experiences are not always encoded as traumatic. Stress is not always a bad thing, after all. From an evolutionary perspective, stress has been necessary for our survival as a species. For example, when faced with a threat, such as a predator entering our village, our internal systems primed us to act quickly – to either fight the threat or escape from the threat (known as the fight-or-flight response). Under normal stress conditions, the body's response is mediated through allostasis, a generally adaptive and temporary process that controls the body's response to a stressor (Sapolsky, 2004). Acute stressors set off a chain reaction that activates the sympathetic nervous system and leads to the secretion of stress hormones, priming an individual to respond to the threat (Sapolsky, 2004). In this type of scenario, it would not have been evolutionarily appropriate or even preferable to be at peace; in fact, it was the very disruption of our peace that enabled us to take actions that helped us to survive. Once the threat was removed, the release of stress hormones would stop and move the body back into a state of balance (Sapolsky, 2004), and we could be at peace once again.

However, in today's modern world, these stressors have taken on a vaguer and more persistent role in our lives. From the around-the-clock news cycle that exposes us constantly to the pains of the world, to hustle work culture that belittles sleep, to a global pandemic, our realities have changed the nature of stress in our lives. And when the stress response is overly activated, whether due to past traumatic experiences or the simple fact of being a human being in today's modern and chaotic world, the system for managing it can break down (McEwen, 1998).

Although individual reactions to extreme stress are not necessarily harmful and can vary from person to person, chronic stress can be toxic and can lead to the disruption of several important psychological and physiological systems, setting off a chain reaction of negative biological, social, and developmental consequences (SAMHSA, 2014; Zarse et al., 2019). These changes can cause a person to be more acutely aware of potential risks within the environment (both real and imagined) and can predispose them to be more sensitive to a range of environmental stimuli, interrupting one's ability to regulate their internal systems into a state in which they can work, learn, and play (van der Kolk, 2015). And while the effects of prolonged and extreme stress have a clear impact on one's own internal sense of peace, it can also have a cascading negative impact on one's ability to form and maintain meaningful relationships with others (van der Kolk, 2015). Maslow's hierarchy of needs (1943) suggests that when basic physiological and safety needs are not met, one cannot move up the ladder, so to speak, to reach higher goals of love and belonging and self-actualization (see Figure 4.1). New research suggests that trauma can even be passed down to future generations through biological changes in gene expression (Yehuda & Lehrner, 2018).

Although it will likely be years, or even decades, until we can better understand the impact that extreme stressors will have on today's generation, many people believe that the need for trauma-informed practice and design that can help people cope with this reality will only increase in the coming years.

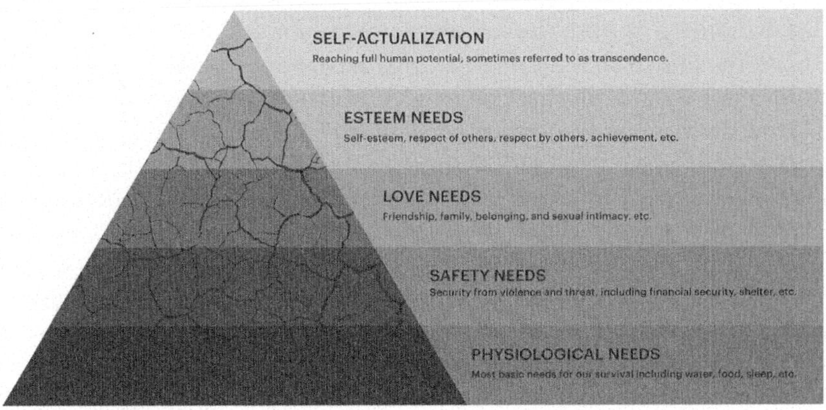

FIGURE 4.1 Maslow's hierarchy of needs. The hierarchy of needs illustrates that higher-order needs can only be met once more basic needs have been satisfied.

Source: Graphic by Erin K. Peavey, HKS Inc., adapted from Maslow (1943).

Trauma-Informed Design

The authors of this chapter, perhaps like many of you, have struggled personally with trauma and have spent years working to cultivate peace both internally and in our external worlds. As we delve into the research that follows, we will focus on the role that design of the built environment can play in creating supportive spaces that help to cultivate peace.

Compared to trauma-informed care (TIC), which describes the approach taken by social services organizations to adequately recognize and respond to the needs of traumatized individuals accessing services (Substance Abuse and Mental Health Services Administration [SAMHSA], 2014), trauma-informed design (TID) aims to apply the principles of trauma-informed care to the built environment, demonstrating how design interventions can best support trauma-informed approaches (TIA) (Farrell, 2018). TIA and TIC are often used synonymously, with TIA encompassing the full range of experiences, services, and spaces that are designed using this lens. The principles of TIA include the following: (1) safety; (2) trustworthiness and transparency; (3) peer support; (4) collaboration and mutuality; (5) empowerment, voice, and choice; and (6) cultural, historical, and gender issues (SAMHSA, 2014). These principles are interdependent and are not meant to be viewed and applied as a checklist but rather should be seen as core considerations in caring for individuals who may have experienced trauma. TID aims to take a similarly holistic approach – going beyond aesthetics to holistically consider how our environments impact our emotional well-being in the face of adversity, and how our internal well-being can impact our connection with others and with the world around us.

Interest in TID has exploded in recent years, a likely byproduct of the rise in the recognition of both the prevalence of trauma and its deleterious effects, as well as the understanding that the built environment can powerfully influence our internal and external experiences. Much of this early work has looked at TID in the context of permanent supportive housing (PSH) facilities and homeless shelters, with the understanding that individuals experiencing homelessness have often experienced significant trauma (e.g. Bollo & Donofrio, 2021; Owen & Crane, 2022). The first known adaptation of TIC principles into TID principles was developed by the Committee on Temporary Shelter (COTS), an advocacy group for individuals experiencing homelessness (Farrell, 2018). Several reports by members of Shopworks Architecture, the Center for Housing and Homelessness Research at the University of Denver, and Group 14 Engineering have provided design firms with practical guidance for applying trauma-informed design principles to their work (Brisson et al., 2023; Grabowska et al., 2021; Wilson et al., 2020).

Others have begun to investigate the application of TID at the neighborhood scale, indicating features like lighting, traffic density, and greenspace as components to consider in a trauma-informed neighborhood (Schroeder et al., 2021). Additionally, a limited number of studies and academic book chapters have evaluated TID through a developmental lens: Kopec and Harte (2020) explored the potential for design to complement trauma-informed approaches in schools, while an earlier example illustrated environmental modifications as a component of a trauma-informed program in a juvenile justice setting (Marrow et al., 2012). Ames and Loebach (2023) explored the potential for TID to support youth in the child welfare system, particularly those in residential treatment settings. Finally, a recent project from the Trauma-Informed Design Society worked with public schools to develop the TiDEvalK12, a TID evaluation tool for K-12 schools (Harte et al., 2022). While research demonstrating evidence for its effectiveness is still in its infancy, it is widely acknowledged by both researchers and practitioners that TID is, at its core, good design and can support the well-being of building occupants.

Building from the trauma-informed principles established by SAMHSA (2014), Figure 4.2 illustrates how these principles, originally designed to support systems of care and operational protocols, can be applied to the design of the built or physical environment to work in concert with trauma-informed programs.

FIGURE 4.2 Inner peace and outer peace go hand in hand, both supporting the other. By utilizing TID principles, we can help to create spaces where people feel safe, calm, and at peace, thus nurturing more peace in the world.

Source: Graphic by Erin K. Peavey, HKS Inc., adapted from SAMHSA (2014).

1. Safety

Creating a feeling of safety and comfort is essential if human beings are to focus on anything outside of their immediate survival. This need is especially pronounced in individuals who have experienced trauma and who may have difficulty feeling safe or regulating their nervous system. The principle of safety not only refers to one's ability to be physically safe but also refers to the psychological mechanisms that help us feel safe, so we are able to move out of survival mode and into a realm of thinking, processing, and moving up the hierarchy of needs. Some evidence-based strategies for designing for comfort and safety include these features:

- **Eliminate triggers:** Reducing and removing environmental stressors and triggers. Known stressors, such as slamming doors or buzzing fluorescent lights, can be mitigated through design strategies.
- **Human scale:** Creating thresholds, niches, and spaces that help people feel safe and comfortable in their space.
- **Biophilia and nature:** Providing opportunities to connect with and view nature and natural elements calms and grounds us. This includes natural light.
- **Movement:** Modern design and culture often prioritize sitting still; however, research shows that rhythmic movement in the form of rocking chairs or fidget items can help soothe rattled nervous systems.
- **Order:** A lack of order can create a sense of lack of safety in our minds. Reducing clutter and chaotic or overly simulating environmental features can help to reduce feelings of overwhelm and assist in nervous system regulation.

2. Trustworthiness and Transparency

The principle of trustworthiness and transparency typically refers to organizational operations and the ways in which decisions are made, with the goal of building and maintaining trust among those receiving and those providing services within an organization. In applying this core principle to the built environment, one can think of the ways in which the environment can engender a sense of trustworthiness and transparency. This overlaps with other principles, such as collaboration and mutuality, due to its relationship with power dynamics. Strategies for addressing this principle in design include:

- **Welcoming:** Provide a welcoming exterior entry that offers a degree of transparency or porosity so people can feel like they know what they are going into and that they are welcome there.
- **Clear wayfinding:** Creating ease of navigation within a building or facility through features like signage or other landmarks can reduce confusion, stress, and anxiety.

- **Visual transparency:** Use of visual transparency and openness, especially in entry spaces or transition spaces, allows people to have a sense of a space before entering, which can reduce fear and anxiety.
- **Comforting spaces:** Warm, soft, and cozy spaces can provide the backdrop for therapeutic and intimate conversations. Use of nature or nature-based imagery can also be comforting.

3. Peer Support

A core feature of good design is the ability to facilitate one of the best coping mechanisms humans have access to: our connection with one another. Our connection with other people, especially peers going through similar experiences, can be grounding and foundational for our sense of safety and belonging in the world. However, there is a growing epidemic of loneliness in the world today, and research has shown that the influence of social relationships (or lack thereof) on mortality risk factors is comparable to known mortality risk factors like smoking and alcohol consumption (Holt-Lunstad et al., 2010). Design can play an important role in helping to create opportunities for social connection (Peavey, 2020; Peavey & Holt-Lunstad, 2021). To provide environments that optimize peer support, design strategies include:

- **Adaptable furnishings:** Flexible layouts that allow for multiple small or large group arrangements, and customization of the space, can help to promote social connection and promote a sense of control over social interactions.
- **Shared space and activities:** Creating shared outdoor and indoor spaces that are designed for mixing and collaborating on different activities (e.g., art, sports, library, marker space) can be a great way to overcome barriers and connect.
- **Equitable provisions:** Avoid emphasizing hierarchical relationships. For example, ensure that those with limited mobility can move around a space as easily as those without by reducing barriers and increasing transition space widths.

4. Collaboration and Mutuality

Collaboration and mutuality are about partnering and leveling power differences between individuals (e.g., staff + patients, teacher + student, administrator + staff, etc.). This recognizes that everyone who is a part of the organization has a role to play in creating a therapeutic and healing experience. Space that helps to flatten the hierarchy between teacher and student, between caregiver and the person(s) being cared for, and between those in power and those who are not can foster a shared sense of mutuality and improved feelings of empowerment and safety for those who may be especially vulnerable. Possible design interventions include:

- **Level the playing field:** Whether through chairs set up in a circle, or sitting on the ground, creating a space where everyone is equal is helpful to support mutuality.
- **Flatten hierarchy:** Create openness and shared spaces for caregivers and care receivers.
- **Accessibility:** Ensure spaces support ease of mobility for all occupants, including those with mobility impairments and other disabling conditions (e.g., compliance with American with Disabilities Act and beyond).
- **Avoid creating barriers:** Avoid creating physical barriers between those receiving and those providing services. While this is often done to protect the safety of staff in environments like inpatient psychiatric facilities, using features like glass partitions, care should be taken to consider alternative measures that can support this goal without furthering us vs. them power dynamics (see Mount Sinai Case Study).

5. Empowerment, Voice, and Choice

Choice is essential in every design, and it plays a special role in empowering those who have been in situations of chronic disempowerment and helplessness due to structural, interpersonal, or cultural issues. By providing choice, we can put the control back in the hands of those who need it most. For example, something as simple as providing a reading lamp or Christmas lights can allow the stark overhead lights to be turned off and enable individuals to choose a cozier feel. For instance, one of the authors of this chapter worked in an office that had a wellness room, featuring dim reading lights and a curtain that covered a bare wall to create a space for calm, relaxation, and emotional reset. Similarly, providing rugs to help create a safe or quiet space featuring soft items can help. This has become standard practice in some daycares, child centers, and schools and could be potentially beneficial in offices across the globe. Creating small opportunities to control one's environment is essential. Some examples of this principle in action include:

- **Flexibility and options:** Allowing for customization of lighting levels (dimmer, task lighting) or providing opportunities for choice over where one sits or locates themselves in relation to others can enable them to feel a measure of control that can help people regulate and feel calm.
- **Calming spaces:** Provide spaces to sit and calm down after a stressful event. For example, some schools feature calm-down corners in classrooms, with soft rugs, pillows, beanbag chairs, and a small warm desk lamp.

6. Culturally Responsive

Across the globe, we have both unique and shared issues around racism, sexism, ablism (bias against people with disabilities), and historical biases. Understanding and being sensitive to these contexts where we are building and designing is extremely important. When spaces are designed in a way that allows individuals to see themselves reflected in the spaces they inhabit – through features like murals, artwork, or space to display personal items – this can support a feeling of anchoring in the space. While this is important in any space, it is especially so in spaces that need to recognize and appropriately honor historic events that shaped a community or legacy (see Sandy Hook Case Study).

- **Cultural symbols:** Lean into symbolic representation of the non-dominant culture or cultural artifacts to signal safety and belonging to members of those cultures in the space, features like a rainbow flag, artwork, or murals from a local artist depicting the local community's people, cultures, struggles, and successes.
- **Sensitivity**: Be thoughtful about design choices that may have implications for varying cultures, beliefs, practices, physical abilities, gender identities, sexual orientation, and so on. For example, consider individual all-gendered accessible bathrooms when possible. If not fully possible, providing an option that serves this function can be helpful.
- **Identity anchors**: People should be able to see themselves represented, literally and metaphorically, in spaces, through interventions like displaying individuals' achievements or artwork or murals that incorporate a breadth of cultural backgrounds. Identity anchors can help to combat isolation by enhancing a sense of belonging in the space.

Case Studies on Trauma-Informed Design

Dr. JoAnn Difede, attending psychologist at NewYork-Presbyterian/Weill Cornell Medical Center, suggests that "as humans, we often think that if the problem seems enormous, the solution has to be, that isn't really the case" (Moog, n.d., para. 3).

On this journey to learn to design for peace, we've discussed why it is important to consider trauma-informed principles and practices and how to apply that framework to the built environment. The final step is to explore examples of these principles applied in the real world to help you to own these ideas as tools to cultivate peace, applying to your own projects and practices.

Designing environments with a trauma-informed lens includes both the *process* of design and the *place* that is created. The process should be one that listens and empowers the voices of diverse stakeholders of the space (e.g., visitors, students, patients, staff, community) rather than what often happens where a very small group of powerful leaders make all of the decisions. The place that is created with the trauma-informed lens should be one that is responsive to occupants' needs for a sense of safety and calm that often exceeds what we consider just 'good design'.

The following case studies are demonstrations of TID in action and help to weave together the different considerations of a trauma-informed approach that can help to promote peace. We chose examples from a variety of populations (e.g., school, psychiatric, housing) and needs (e.g., healing, recovery, growth) to help provide a diverse perspective that can be applied to a range of needs.

Case Study 1: Sandy Hook Elementary School

Location: Newtown, Connecticut
Firm: Svigals + Partners

There are few places where a trauma-informed approach to design would appear more relevant and necessary than in the redesign of Sandy Hook Elementary school, the site where twenty children between six and seven years old and six teachers were killed in a deadly school shooting in 2012. Following the massacre, the local community voted to demolish the school building and design a new school that would support the community as they began the painful process of healing. How do you design a building that both honors the victims of a tragedy while helping those who survived and future generations move forward with some sense of normalcy?

Safety and Security

As alluded to in earlier portions of this chapter, one of the primary features of a trauma-informed building is an emphasis on safety and security, both real and perceived, physical and emotional (see Figure 4.3). One overarching approach that was taken to address security at the new school draws on a concept called Crime Prevention Through Environmental Design (CPTED), which emphasizes clear sightlines and openness to support security goals. By setting the building further back on the original property, the building is not as easily detectable from the main road. The new building also sits on a higher elevation, ensuring that administrators can see people approaching the school. Some security features are disguised, such as the bulletproof glass used in all the windows and classroom doors that lock and appear to be made of wood but are constructed from heavy stainless steel.

FIGURE 4.3 Exterior façade of Sandy Hook Elementary School.

Source: Robert Benson Photography, Svigals + Partners Architecture + Art.

While physical security and safety were primary concerns in the redesign of Sandy Hook, the school also needed to promote learning and creativity. One strategy was to provide built-in 'treehouses,' offering retreat spaces for students and serving as spaces for small group gatherings. Acoustic panels shaped like tree leaves line the interior of the building and serve the dual purpose of tying the design to the natural environment and absorbing potentially triggering loud noises (see Figure 4.4). Although research on the impact of color on emotion is limited, metal blinds in the atrium lobby that cast colors on the interior spaces and strong colors in the furnishings may promote a sense of cheerfulness (see Figure 4.5).

FIGURE 4.4 Interior treehouse at Sandy Hook Elementary School.

Source: Robert Benson Photography, Svigals + Partners Architecture + Art.

FIGURE 4.5 Atrium lobby at Sandy Hook Elementary School.

Source: Robert Benson Photography, Svigals + Partners Architecture + Art.

Empowerment, Voice, and Choice

Teachers, parents, children, and local community members engaged in the design process. Participants shared their thoughts and feelings about what the school meant to them and how it should be designed to best represent the needs and desires of the local community. Children were invited to create artwork to be showcased in the new school and were involved in participatory design workshops, where the contractors talked to them about the new construction.

Cultural, Historical, and Gender Issues

Local community members emphasized the town's strong connection to nature, so incorporating the local environment, natural materials, ample daylight, and outdoor courtyards was critical for the project. The structure features an undulating, wave-like facade made from two types of hardwood. Three footbridges, a feature found within the local community, were incorporated into the new school's entrance. These features also add an additional layer of security, requiring all visitors to enter the building through three highly surveilled pathways. The new facility features three open courtyards and ducks, which were part of the previous school and are incorporated into artwork and murals throughout (see Figure 4.6).

FIGURE 4.6 Interior wall at Sandy Hook Elementary School.

Source: Robert Benson Photography, Svigals + Partners Architecture + Art.

For the source of this information and more details on this case study, see Campisi (2018), Jacobs (2016), Steinberg (2016), and Urist (2014).

Case Study 2: Mount Sinai Beth Israel Comprehensive Behavioral Health Center

Location: New York, NY
Firm: HKS

Mount Sinai's new Comprehensive Behavioral Health Center offers an integrated continuum of services for patients, from inpatient and outpatient mental health programs to substance use disorder treatment and primary care. Many similar facilities have these programs located in separate buildings or locations, which can be both an inconvenience and a barrier to accessing care. While the main hospital is located uptown, the new behavioral health facility was built downtown adjacent to the communities in which many patients they live and work. The new facility cost $140 million to build – believed to be the largest private investment in mental health care in New York state history.

Although the principles of TID, as we now understand them, were incorporated into the design of the facility, TID goals were not discussed using that terminology. However, the impetus behind the new comprehensive behavioral health facility and the final output of the design aligns with the principles that have been mentioned in this chapter – further emphasizing the point that TID is, at its core, good design.

Safety and Security

In a behavioral health facility, safety is of critical importance. The ability for staff to be able to monitor patients on the unit is one of the primary needs of patients requiring this level of care. Centrally locating the nurse stations in multiple locations throughout the unit helps to improve visibility of patients as they move about the space throughout the day. Low glazed barriers along the perimeter of the nurse station allow patients to communicate with staff members while also restricting access to risk items that might be located within the nurse station (see Figure 4.7). This design feature also serves to reduce feelings of uneven power dynamics that can be elicited when offices and spaces are closed off behind full-length walls, glass partitions, and closed doors.

Furniture located throughout the unit and in patient rooms was selected to ensure patients' safety, by considering weight, form, and configuration. The angles of certain walls in patient rooms were set to maximize visibility and reduce blind corners while also enabling patients to have some privacy.

FIGURE 4.7 Nurse station at Mount Sinai Beth Israel Comprehensive Behavioral Health Center.

Source: Garrett Rowland Photography.

Peer Support and Collaboration

Providing options that allow patients to moderate their level of social interaction with others is critical in acute mental and behavioral health settings, as it may offer patients' a sense of control in an extremely controlling environment. In the Comprehensive Behavioral Health Center, gathering spaces and open hubs are located throughout the facility to provide patients with options for social connection (see Figure 4.7). A mix of seating options, moveable furniture, and noise-reducing barriers offer additional choices to patients. These spaces can also double as locations for collaboration among patients and members of their treatment team.

Trust and Transparency

Through the design of a facility, organizations can potentially communicate that they can be trusted with their care. The entry of the new Comprehensive Behavioral Health Center, featuring tall ceilings, natural materials, and earth tones, contrasts with other mental and behavioral health facilities, where heavy metal doors and long, dark hallways often mark the entrances, an approach that might engender comfort and confidence (see Figure 4.8). A mission of the organization is to treat all patients with dignity and respect. The inclusion of a small, shared desk for patients directly adjacent to the nurse station creates a neutral space, not owned by any party but rather a shared space for conversation that helps reduce perceived power imbalances.

FIGURE 4.8 Entrance at Mount Sinai Beth Israel Comprehensive Behavioral Health Center.

Source: Garrett Rowland Photography.

In the new behavioral health center, the long hallways are broken up with colors, patterns, and nature artwork. Benches located in the hallways serve as retreat spaces, where patients can go to be alone or have a private conversation with visitors, treatment providers, or peers (see Figure 4.9).

FIGURE 4.9 Transition space at Mount Sinai Beth Israel Comprehensive Behavioral Health Center.

Source: Garrett Rowland Photography.

For the source of this information and more details, see HKS (2021).

Case Study 3: Arroyo Village

Location: Denver, Colorado
Firm: Shopworks Architecture

Unstable housing can be a source of trauma. Homeless and houseless individuals often experience high levels of past trauma exposure and have a higher occurrence of co-occurring conditions, such as mental health and substance abuse disorders. Given this, TID principles should be considered for any space that serves this group of individuals.

Arroyo Village is a development located in Denver, Colorado, providing housing services, ranging from a homeless shelter to permanent supportive housing (PSH) and affordable housing units. Arroyo Village was designed explicitly using TID principles to with the intention of addressing the needs of residents. In addition to housing, counseling and additional support services are incorporated.

Safety and Security

Having visual and physical access to nature, natural materials, and nature images may enhance a person's sense of psychological safety. A central courtyard at Arroyo Village includes a space for gathering and gardening (see Figure 4.10). Interior spaces surrounding the courtyard allow natural light to enter the building through wide banks of windows. By maintaining views into the courtyard, principles of prospect and refuge are acknowledged. Wood materials were also incorporated into both interior and exterior spaces to further enhance the biophilic design (see Figure 4.11).

Noise is one of the most common environmental triggers for trauma responses, so the housing units at Arroyo Village were designed to ensure that the individual units were adequately soundproofed. For those who have experienced homelessness in particular, a sense of security over personal belongings is also important. By providing a place to store personal belongings in lockers that can be locked, Arroyo Village may enhance residents' sense of security and control.

FIGURE 4.10 Courtyard and garden at Arroyo Village.

Source: Ben Eyster Photography.

FIGURE 4.11 Interior space incorporating natural elements at Arroyo Village.

Source: Ben Eyster Photography.

Cultural, Historical, and Gender Issues

In recognition of a long-standing tradition that the original shelter had of providing guests hand-made quilts made by volunteers, the quilt motif was used as inspiration for design elements of the new development, such as the patterned exterior façade (see Figure 4.12).

FIGURE 4.12 Exterior façade at Arroyo Village.

Source: Matthew Staver Photography.

For the source of this information and more details, see Grabowska et al. (2021), Kimura (2022), and Rodgers (2021).

Conclusion

In light of our growing understanding of both the prevalence of trauma and its implications for the current global mental health crisis, designers must take TID into account when designing facilities that aim to support individuals who have experienced stressful and traumatic experiences. Throughout this chapter, we have introduced trauma-informed design as an approach to promoting both internal and external peace. While this is an evolving field and empirical evidence for its appropriate application and effectiveness is still emerging, we have provided best practice examples that can be used to inform future design work.

At the end of this chapter, you will find a tool that can be used to support you and your team as you engage in this work (see Table 4.1). This tool, drawn from trauma-informed care principles detailed in the SAMHSA report (2014) and TID principles featured elsewhere in this chapter (e.g. Farrell, 2018), will

TABLE 4.1 Peace- and trauma-informed design exercise. Do you want to create a project that supports peace and kindness and is sensitive to those who have been affected by trauma? Use this table to test your project or initiative for how it supports peace through a trauma-informed lens using this exercise.[1]

Trauma-informed principles (TIP)	*Physical environment (Physical environment attributes to promote peace, following the TIP framework)*	*Programs (Program attributes to promote peace, following the TIP framework)*	*Policy (Policy attributes to promote peace, following the TIP framework)*
Safety Person(s) feels physically and psychologically safe.	Does the space have blind corners or dark spots in hallways or paths of travel? (Y/N) – If so, consider how to eliminate those through adding lighting, furniture placement, or other means. Does the space provide an opportunity to get away to a reduced-stimulation area if needed? (Y/N) – If not, consider how to add a spot with softer furnishing and warmer light that is a little removed from the action. Does the space provide opportunities to connect with and view nature and natural elements? (Y/N) Does the space help to reduce clutter/have convenient places to put things away after use, so that it is not visually chaotic? (Y/N) Does the space and/or furnishings allow opportunities for movement (e.g., rocking chairs, swivel, standing desks)? (Y/N)	Do the programs provide opportunities for both the more reserved, quiet members as well as the more vocal, extroverted members? (Y/N) Do the program's leaders create a space where people are free to speak their minds and share their experiences? (Y/N)	Does the policy support physical safety for all members (e.g., staff, clients, allied members)? Does the policy support psychological safety for all members (e.g., staff, clients, allied members)?

(*Continued*)

HumanWindow. (2019, June 14). Dr Gabor Maté on childhood trauma, the real cause of anxiety, our 'insane' culture and ayahuasca [Video]. *YouTube*. https://www.youtube.com/watch?v=e7pV0IPWUlI

Jacobs, K. (2016, August 23). Sandy Hook rebuilds. *Architect Magazine*. https://www.architectmagazine.com/design/sandy-hook-rebuilds_o

Kimura, D. (2022, August 8). How trauma-informed design is incorporated into supportive housing: Industry experts share best practices for creating an environment where residents can feel safe and thrive. *Affordable Housing Finance*. https://www.housingfinance.com/developments/how-trauma-informed-design-is-incorporated-into-supportive-housing_o

Kopec, D., & Harte, J. D. (2020). Design as the missing variable in trauma-informed schools. In E. Rossen (Ed.), *Supporting and educating traumatized students* (pp. 343–358). Oxford University Press. https://doi.org/10.1093/med-psych/9780190052737.003.0019

Marrow, M. T., Knudsen, K. J., Olafson, E., & Bucher, S. E. (2012). The value of implementing TARGET within a trauma-informed juvenile justice setting. *Journal of Child & Adolescent Trauma*, *5*(3), 257–270. https://doi.org/10.1080/19361521.2012.697105

Maslow, A. (1943). A theory of human motivation. *Psychological Review, 50*(4), 370–396. https://doi.org/10.1037/h0054346

McEwen, B. S. (1998). Protective and damaging effects of stress mediators. *New England Journal of Medicine, 338*(3), 171–179. https://doi.org/10.1056/NEJM199801153380307

Moog, M. (n.d.). Understanding PTSD in uncertain times. *Health Matters: Stories of Science, Care & Wellness and New York-Presbyterian*. https://healthmatters.nyp.org/understanding-pandemic-related-ptsd/#:~:text=More%20than%20two%20years%20into,called%20post%2DCOVID%20stress%20disorder

Núñez-González, S., Delgado-Ron, J. A., Gault, C., Lara-Vinueza, A., Calle-Celi, D., Porreca, R., & Simancas-Racines, D. (2020). Overview of "systematic reviews" of the built environment's effects on mental health. *Journal of Environmental and Public Health, 2020*. http://doi.org/10.1155/2020/9523127

Owen, C., & Crane, J. (2022). Trauma-informed design of supported housing: A scoping review through the lens of neuroscience. *International Journal of Environmental Research and Public Health, 19*(21), 14279. https://doi.org/10.3390/ijerph192114279

Peavey, E. (2020). Connecting IRL: How the built environment can foster social health. *HKS Architects*. https://www.hksinc.com/media/7347/connecting_irl_hks_white_paper.pdf

Peavey, E. (2023). How school design can help children feel safe: Trauma-sensitive design can create safety and comfort, essential for learning. *Psychology Today*. https://www.psychologytoday.com/us/blog/designed-for-happiness/202308/how-school-design-can-help-children-feel-safe

Peavey, E., & Holt-Lunstad, J. (2021, May 3). Is your environment making you lonely? *Psychology Today*. https://www.psychologytoday.com/us/blog/designed-happiness/202105/is-your-environment-making-you-lonely

Rodgers, J. (2021, April 23). Arroyo village – A home for healing. *Colorado Builder*. https://coloradobuildermag.com/design/arroyo-village-a-home-for-healing/

Sapolsky, R. M. (2004). *Why Zebras don't get ulcers* (3rd ed.). Holt Paperbacks.

Schroeder, K., Noll, J. G., Henry, K. A., Suglia, S. F., & Sarwer, D. B. (2021). Trauma-informed neighborhoods: Making the built environment trauma-informed. *Preventive Medicine Reports, 23*, 101501. https://doi.org/10.1016/j.pmedr.2021.101501

Steinberg, A. (2016, September 16). Can you erase the trauma from a place like Sandy Hook? *The New York Times*. https://www.nytimes.com/2016/09/15/magazine/can-you-erase-the-trauma-from-a-place-like-sandy-hook.html

Substance Abuse and Mental Health Services Administration. (2014). *SAMHSA's concept of trauma and guidance for a trauma-informed approach*. HHS Publication No. (SMA) 14–4884. Substance Abuse and Mental Health Services Administration. https://store.samhsa.gov/sites/default/files/d7/priv/sma14-4884.pdf

Urist, J. (2014, November 20). The architecture of loss: How to redesign after a school shooting. *The Atlantic*. https://www.theatlantic.com/education/archive/2014/11/the-architecture-of-loss-how-to-redesign-after-a-school-shooting/382952/

Van der Kolk, B. (2015). *The body keeps the score: Brain, mind, and body in the healing of trauma*. Penguin Books.

Wilson, J. Macur, R. Rossbert, L., & Holtzinger, C. (2020). *Designing for healing, dignity, and joy*. Shopworks Architecture, University of Denver Center for Housing and Homelessness Research, & Group14 Engineering. https://shopworksarc.com/wp-content/uploads/2020/06/Designing_Healing_Dignity.pdf

World Health Organization. (2022). *World mental health report: Transforming mental health for all*. https://www.who.int/publications/i/item/9789240049338

Yehuda, R., & Lehrner, A. (2018). Intergenerational transmission of trauma effects: Putative role of epigenetic mechanisms. *World Psychiatry, 17*(3), 243–257. https://doi.org/10.1002/wps.20568

Zarse, E. M., Neff, M. R., Yoder, R., Hulvershorn, L., Chambers, J. E., & Chambers, R. A. (2019). The adverse childhood experiences questionnaire: Two decades of research on childhood trauma as a primary cause of adult mental illness, addiction, and medical diseases. *Cogent Medicine*, *6*(1), 1581447. https://doi.org/10.1080/2331205X.2019.1581447

OTHER VOICES

Mardelle McCuskey Shepley

When writing about peace, one is committed to provide a podium for as many voices as possible. The following essays offer an abbreviated platform for both design practitioners and academics who are conducting work related to peace by design. The cameo authors highlighted in this section are multinational and represent the fields of architecture, urban design, nursing, fashion apparel, environmental psychology, anthropology, education, and philosophy. Their one-sentence bios attempt to honor their multidisciplinary approaches.

Design researcher/educator/philosopher Renato Troncon and healthcare designer/patient advocate/educator Tammy Thompson set the stage for subsequent cameos by providing richly distinct approaches yet converging perspectives on the power of design. Troncon reminds us of the "inseparability in doing and thinking" when seeking value in architecture and design, while Thompson advocates for experiences and environments that support belonging and "allow each person their own ethnographic discovery."

DOI: 10.4324/9781003462019-5

Architectural designer/planner/community advocate Taft Cleveland bears witness to urban food inequities and racial disparities reminding us that "racial health inequity is a matter of life or death!" Environmental psychologist educator/programmer Lorraine E. Maxwell moves in a similar vein, reminding us that "ensuring social justice in our society requires multifaceted approaches" and describes the multiple ways in which there is discriminatory access to public physical environments.

Designer/researcher/author Connie Lin and architect/educator/author Arsenio T. Rodrigues have both conducted research on environments that support personal peace. Lin describes related research and notes "the presence of water ranging from a rippling lake or trickling fountain" as a source of stress reduction, while Rodrigues shares his vision of sacred space and peace via "archetypes that contribute to sacred place creation."

Design researcher/nurse/ educator Marie Elf emphasizes the contribution of the environment to rehabilitation and recovery. "When consciously adapted, [it] has the potential to be a powerful link in a person's recovery." Architect/researcher/

neuroscientist Evangelia Chrysikou reminds us of the importance of privacy, particularly in sexual healthcare spaces, one of the most stigmatized areas of health. Fashion designer/educator/anthropologist Denise Nicole Green shares her conviction that peace in the field of fashion design "involves embracing, seeking, and making space for seemingly divergent forces."

We invite our readers to reach out to the editor to share their perspectives as well.

ONLY THOUGHTS AND PRAYERS?

Renato Troncon

In the summer of 2023, the Procuratie Vecchie in Piazza San Marco hosted an exhibition titled "A World of Potential." The event, initiated by The Human Safety Net Foundation, was clearly inspired by Maria Montessori's 1948 essay "To Educate the Human Potential" (Montessori, 2015). The exhibition offered an immersive and interactive experience in human virtues, character, and social strengths useful in mastering the many critical situations that cloud the sense of humanity. These principles of positive psychology included creativity, perseverance, gratitude, curiosity, hope, social intelligence, teamwork, empathy, and courage. They were translated into 16 interactive multimedia installations, both analog and digital, with a thousand books distributed on open shelves with a direct view of the Square.

The exhibition is a highly motivational oasis of reflection. If the bibliographic lineup of psychology, neuroscience, essays on ethics, political criticism, and pedagogy correctly advises us to consider that "social intelligence," "teamwork," and "empathy" are at the heart of all positivity, the passionate visitor asks themselves, what is their "technique" and what are their tools for their promotion and "enabling?"

Or are they "only thoughts and prayers?"

This question is central to the discussion of how design and architecture can promote peace. As Kant wrote in "Perpetual Peace," the natural state of humanity is war, not peace (1795). But if this is the case, we must ask ourselves with much more seriousness how we go about building it.

Do we know an "Art of Peace" as we know Sun Tzu and his "Art of War"?

DOI: 10.4324/9781003462019-6

Here, in drafting an "Art of Peace," architecture and design must set goals that are not only practical and poetic but also strategic.

Indeed, what is happening in contemporary architecture resembles, in its best parts, a large and material exemplification of vast parts of a living and concrete pedagogy, a teaching of how to build community. Entire sectors are flourishing, from architecture for educational spaces to architecture for healthcare spaces and urban recovery interventions. In these sectors, projects are advancing with results intended to support multiple life circumstances: inclusive spaces, spaces for participation and facilitating dialogue, spaces that promote a sense of security and strengthen identity and culture, spaces for care and intergenerational encounters, spaces for artistic expression, and spaces consecrated to memory. Design can contribute to developing media campaigns, social responsibility, and social entrepreneurship.

But it is equally true – and indeed even more so – that in this way architects and designers fulfill a conceptual and psychological function, bringing life to the "transformative thinking" that unfolds along a sequence of values. Not in a generically ethical sense but in a professional logic sense, even in relatively less complicated projects, we are confronted with the complexity made up of stratifications and hidden dimensions and entanglements and conflicts, which may offer solutions and opportunities. Louis H. Sullivan, in his rationalistic approach, wrote for this reason that much of the architect's work is to discover the "functions" to which a "form" must respond (1924). Buckminster Fuller declared that designers must grow in a "comprehensive" curriculum, which allows them to take broad initiatives "from economics to science, to industry, to technologies" to make the world more peaceful, prosperous, and egalitarian (1963). Christopher Alexander assigned to architecture the task of intertwining the basic elements of living in a variety of "vital patterns" (1979).

It is this "value-based" approach that characterizes good architecture and design. And so, what role do they have in putting what is positive – "social intelligence," "teamwork," "empathy" – at the service of building peace? I believe that, for this purpose, we must go beyond the dissemination of design and architecture tools as it happens today, separated from their inspiration. Instead, we must carry, into the strategic processes of peacebuilding, the inseparability of doing and thinking that characterizes architecture and design. This means valorizing aesthetic techniques. Even simple spontaneous performances, which foster the aesthetic reappropriation of spaces and objects, help integrate the creativity of forms with the reality of functions (Troncon, 2019). (See Figure E.1.) I believe, finally, that only in this way can the communities of architects and designers put their mindsets first, and then their more "technical" design tools at the service of "confluence of disciplines, practices, and expertise." We must rely on these tools if we want to change and heal the world as peacebuilding requires (Lederach et al., 2010).

FIGURE E1.1 Even simple spontaneous performances, which foster the aesthetic reappropriation of spaces and objects, help integrate the creativity of forms with the reality of functions.

Source: "Coppia che danza, Piazza della Libertà, Salerno," Photo credit: Luca Chistè, Phf Photoforma, Trento.

The next installation at the Procuratie Vecchie could thus be dedicated to "how" to prevent the spread of the terrible wounds and scars that wars are causing even now as we write these lines and that, more than ever, are increasing.

It's true. Any war is always a defeat, but architecture and design know how to organize the best energies and characters to make it understood.

References

Alexander, C. (1979). *The timeless way of building*. Oxford University Press.

Fuller, R. B. (1963). *Utopia or oblivion: The prospects for humanity*. Estate of R. Buckminster Fuller.

Kant, I. (1795). Toward perpetual peace. In I. Kant (Ed.), *Theories of federalism: A reader* (pp. 87–99). Palgrave Macmillan US.

Lederach, J. P., Appleby, R. S., Philpott, D., & Powers, G. F. (2010). Strategies of peace: Transforming conflict in a violent world. In *Strategic peacebuilding: An overview* (pp. 19–44). Oxford University Press.

Montessori, M. (2015). *To educate the human potential*. Ravenio Books.

Sullivan, L. (1924). *The autobiography of an idea*. Press of the American Institute of Architects.

Troncon, R. (2019). The perspective of "festiveness" as a model of a new aesthetics. *Castelli di Yale*, *7*, 243–250. https://doi.org/10.15160/2282-5460/2182

I BELONG

We Belong Together

Tammy Thompson

Where do I belong? As an African American female architect – a group that represents not even one percent of the total architects in America – I feel compelled to persevere in a profession where I do not always feel welcome. If not for the growing number of minority aspiring architects I have taught, then for the millions of vulnerable stakeholders I have considered when designing health care experiences, I will continue to design.

I transitioned from my role as a research architect in a health care design firm to focus my work on the education of other designers who may lack lived experience as hospital patients. I recognized a plethora of unmet needs that I had not only heard from patients, and learned from research, but had personally experienced as a patient with a chronic condition distinct to people of color.

As a leader and design professional, I wondered why my experiences had not been similar to those of my Caucasian counterparts. I have been held at police gunpoint under suspicion of intruding into my own place of business. I have had my life and the life of my baby placed at risk by my obstetrician, who failed to make himself available to deliver my baby, because he did not believe me when I told him I was in labor. Despite my education and economic status, racial biases still placed me in vulnerable, life-threatening situations.

Maternal mortality continues to rise in the US, despite its decline across the world. Of these tragedies, African American deaths exceed Caucasian deaths nearly three to one (Saluja & Bryant, 2021) even though Caucasian American women of childbearing age outnumber African American

DOI: 10.4324/9781003462019-7

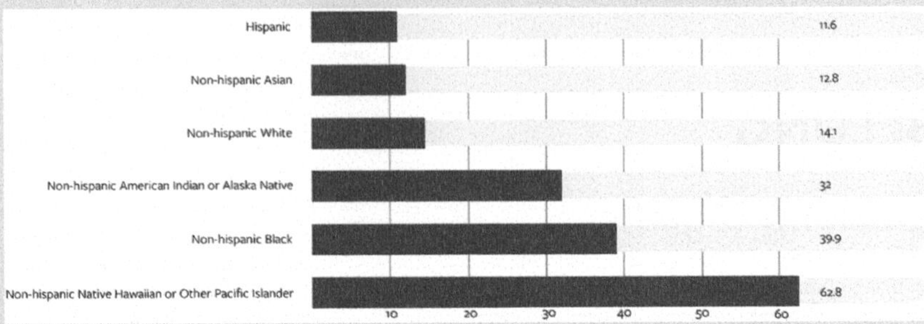

FIGURE E2.1 Pregnancy-related morality (ratio per 1,000 births).

Source: Centers for Disease Control and Prevention (2023).

women of childbearing age nearly four to one (March of Dimes, 2022) (see Figure E2.1). Racism, manifested in numerous forms, is believed to be the cause of many of these disparities, which affect African American women of any social status (Villarosa, 2022).

Where do I belong? I find myself at the intersection of racial bias, gender bias, and ability bias, as a patient with physical limitations caused by my chronic illness. This affects my career, health, safety, and morale. However, I belong to a growing number of design activists who are ready for change. This group transcends gender, age, race, and any other silos built by a system that tries to separate us. Together, there is a place for diverse thinkers to facilitate peace for every individual with design that will elevate communities to achieve equity and heal harms. My privilege as an educator and an architect is to allow my lived experience to fuel my passion for change and to inspire a new generation of limitless thinkers to action. Creative brains hold the solutions to today's complex challenges (Pink, 2012).

How might we design a more peaceful and harmonious future? We can start by planning more diverse neighborhoods. Jane Jacobs (1993) emphasized the value of community, eyes on the street, and an invitation for organic interactions that both feed a sense of community and nurture the melting pot that contributes to a thriving city.

We can promote more collaboration and inclusion in the spaces we create, through co-design with diverse stakeholders. Whether planning schools in neighborhoods that celebrate cultural diversity or creating access to safe playgrounds, fresh food, transportation to work, and health care, design can create pathways to resources for all communities.

Design should facilitate learning, break barriers, and celebrate truth. Like an intricate mosaic, design can highlight the value of each role, culture, and personality, grounded in a parti that celebrates their commonalities. It is only when people come together that they can learn from their own experiences with one another and differentiate the truth from stigmas. In an age of false narratives, designers must create spaces that allow each person their own ethnographic discovery from the diversity around them. As the world recovers from harms caused by the social isolation of a global pandemic, design should promote interpersonal peace by re-introducing individuals to each other and creating safe spaces for social growth, understanding, appreciation, and healing.

References

Centers for Disease Control and Prevention. (2023). *Pregnancy mortality surveillance system*. Centers for Disease Control and Prevention: Reproductive Health. https://www.cdc.gov/maternal-mortality/php/pregnancy-mortality-surveillance/index.html

Jacobs, J. (1993). *The death and life of great American cities*. Vintage.

March of Dimes. (2022). *Population of women 15–44 years by Race/ethnicity: United States, 2021*. March of Dimes PeriStats: Population. https://www.marchofdimes.org/peristats/data?reg=99&top=14&stop=127&lev=1&slev=4&obj=3

Pink, D. H. (2012). *A whole new mind: Why right-brainers will rule the future*. Penguin.

Saluja, B., & Bryant, Z. (2021). How implicit bias contributes to racial disparities in maternal morbidity and mortality in the United States. *Journal of Women's Health*, *30*(2), 270–273. https://doi.org/10.1089/jwh.2020.8874

Villarosa, L. (2022). *Under the skin: The hidden toll of racism on American lives and on the health of our nation*. Anchor.

TRANSFORMING FOOD APARTHEID INTO COMMUNITY HEALTH AND RACIAL EQUITY

Taft Cleveland

The Problem: Food Apartheid

As a child, I experienced asthma. My mother and I (Figure E3.1) would leave the neighborhood hospital in Detroit and go to expensive neighborhood grocery stores that reeked from food spoilage. The other choice was a dollar meal from a fast-food restaurant consisting of a greasy burger and fries accompanied by a high calorie beverage. This is commonplace in poverty-stricken communities (Corcoran, 2021; Gripper et al., 2022).

These residents grapple with what is often termed "food apartheid," a concept popularized by food justice advocate Karen Washington. This term highlights the disparities of whole food systems in Black neighborhoods that are deeply rooted in the systemic injustices of redlining, urban planning, disinvestment, and zoning.

Access to fresh, affordable groceries intersects with racial, geographical, and economic disparities perpetuating structural racial injustices. This cycle also results in generations facing health conditions like obesity, diabetes, hypertension, kidney disease, heart disease, and cancer, primarily affecting Black communities.

Racial health inequity is a matter of life and death! Fifty-six years ago, Dr. Martin Luther King Jr. said, "Injustice in health is the most shocking and the most inhuman because it often results in physical death" (1966). In other words, Black lives end because of the injustice of poverty, racism, insufficient education, and housing, which are the social determinants of health. One strategy for addressing the correlation

DOI: 10.4324/9781003462019-8

FIGURE E3.1 A mother-son journey navigating food apartheid.

between food apartheid and health wellness is the Community Achievement Racial Equity (C.A.R.E.) framework.

The Project: The C.A.R.E. Framework

As a healthcare planner focused on health environments and urban planning, I personally resided in an under-resourced community that suffered from food apartheid and disparities in access to healthcare. I consider this a matter of life and death, and so I designed the Community Achievement Racial Equity framework, known as C.A.R.E.

The C.A.R.E. framework identifies existing conditions, offers recommendations based on proven research and engagement efforts, creates a food justice ecosystem of stakeholders to develop their own rating systems to close the gap on racial health equity, and collaborates on hospital goals with the mission to be a good neighbor to the community they serve.

Preliminary Research

At the onset of the project, I consulted with nutrition experts from the University IL-Chicago, Loyola University, Wayne State University

Center of Urban Studies, the Detroit Black Community Food Security Network, and the Detroit Food Policy Council. Next, I used GIS mapping to identify food apartheid, food oases, and racial disparities surrounding two major hospitals in Detroit, DMC Sinai and Henry Ford Hospital. This data will be shared with a food justice ecosystem composed of city leadership, advocacy groups, policy makers, community residents and stakeholders, urban and health planners, economic developers, and hospitals. They will work in collaboration to develop equitable strategies that will close the racial disparity gaps found in food apartheid and identify the kind of good neighbor a hospital should be.

Based on my findings, I created the C.A.R.E. framework to provide a strategic plan to ascertain the cause of food apartheid in the physical environment and develop solutions to help communities end food insecurity. C.A.R.E. consists of five major components: map disparities, foster innovative and collective responsibility, establish priorities, promote healthier communities, and engage civic leadership (see Figure E3.2).

Map Disparities

I researched the environmental factors impacting Black neighborhoods, focusing on redlining, city design, lack of investment, and zoning to identify community disparities in food access. My data also identified the physical, economic, and social barriers that affect health wellness.

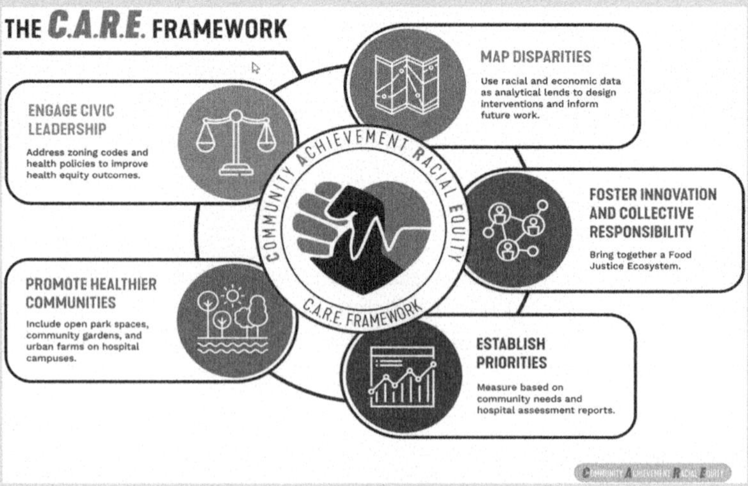

FIGURE E3.2 The Community Achievement Racial Equity framework tenets for wellness.

Foster Innovation and Collective Responsibility

The food justice ecosystem composed of city leadership, advocacy groups, policy makers, community residents and stakeholders, urban planners, economic developers, and hospitals will work in collaboration to develop equitable strategies that will close the racial disparity gaps found in food apartheid neighborhoods.

Establish Priorities

According to the community health needs assessment and data, one of the top priorities to improve the quality of life for residents is access to healthy food. The aforementioned nutrition experts will use the C.A.R.E. framework to develop doable strategies within the radius of hospitals.

Promote Healthier Communities

C.A.R.E. recommendations include developing open parks, bike lanes, community gardens and urban farms, and affordable food stations on hospital campuses.

Engage Civic Leadership

C.A.R.E. delivers a bottom-up approach strategy to empower the collaboration of community partners, policymakers, stakeholders, city leadership, and hospital administrators at the table to improve health outcomes.

The Promotion: Racial Equity, Community Empowerment, and Peace

Dr. Martin Luther King Jr. said, "true peace is not merely the absence of tension, it is the presence of justice" (1958). This understanding recognizes historical injustices as interconnected elements crucial to the pursuit of peace and harmony within marginalized communities.

Just as King identified true peace with justice, the C.A.R.E. framework recognizes that a significant shift must be made towards food justice in Black neighborhoods. Its movement is propelled on understanding, peace promotion, and human upliftment.

As I reflect upon my childhood struggles to access affordable fresh food, I now envision a community abundant with accessible produce, emblematic of the positive change led by food advocates in Detroit. This transformation fosters a tranquil environment of peace and unity, intertwining the principles of food justice within the fabric of community harmony.

References

Corcoran, M. P. (2021). Beyond 'food apartheid': Civil society and the politicization of hunger in New Haven, Connecticut. *Urban Agriculture & Regional Food Systems*, *6*(1), e20013. https://doi.org/10.1002/uar2.20013

Gripper, A. B., Nethery, R., Cowger, T. L., White, M., Kawachi, I., & Adamkiewicz, G. (2022). Community solutions to food apartheid: A spatial analysis of community food-growing spaces and neighborhood demographics in Philadelphia. *Social Science & Medicine*, *310*, 115221. https://doi.org/10.1016/j.socscimed.2022.115221

King, M. L. (1958). *Stride toward freedom.* Harper and Brothers.

King, M. L. (1966, March 25). Presentation at the annual meeting of the medical committee for human rights. Chicago, IL. Quoted in Martin, M. L., Snyder, A. E., Martin, J. M., Skylark, T. S., Firdyiwek, H., & Walker, L. (2017). African American healthcare: Assessing progress and needs through Martin Luther King's perspective on social justice and equality. In M. L. Clemons, D. L. Brown, & W. H. L Dorsey (Eds.), *Dream and legacy: Dr. Martin Luther King in the post-civil rights era* (pp. 59–80). University Press of Mississippi.

SOCIAL JUSTICE/PUBLIC PLACE

Lorraine E. Maxwell

What role does design have in social justice? How do designers influence peace in our society? The design of the physical environment, especially in the public arena, regardless of scale, reflects society's values. If groups of people are left out of the public arena, justice is not served and certainly societal peace is compromised. Therefore, design, and designers, have a prominent role to play in furthering social justice goals leading to peace.

It is not always obvious who a space or product is intended for. Conversely, it may not be obvious who was not considered as a potential user of a space or product. For example, who were the anticipated customers of a restaurant that places the public restrooms accessible only via a flight of stairs? Who were the potential students in a classroom so complex and without spaces for respite that sensory overload was inevitable? Who were anticipated users of a public building that requires no stairs to enter, doors that open automatically whenever anyone approaches, and stairs and elevators both close by and attractively designed?

Consider this: prior to the passage of the ADA, anyone who was a wheelchair user was limited to participation in many public spaces due to the inability to enter the space or use public restrooms. In the 1970s and 80s when shopping malls became major social gathering spaces for teenagers, a young person using a wheelchair probably could not participate in this activity because public restrooms accommodating wheelchair users were not available. That young person couldn't even go to the movies with friends. The ADA, while not perfect or all inclusive, opened employment, public places, and transportation venues to persons with a range of physical disabilities,

DOI: 10.4324/9781003462019-9

FIGURE E4.1 Access to ramp in a wheelchair.

Source: "Nurul Muttaqin Mosque – Wheelchair Ramp" (https://commons.wikimedia.org/w/index.php?curid=75883311) by Chongkian (https://commons.wikimedia.org/wiki/User:Chongkian) is licensed under CC-BY-SA-4.0 (https://commons.wikimedia.org/wiki/Category:CC-BY-SA-4.0), via Wikimedia Commons.

making it possible for a larger variety of people to feel welcomed in these spaces and to participate in public life (see Figure E4.1).

The ADA addressed the needs of persons with a wide range of disabilities. Signs indicating *Whites only* or *Colored only* on water fountains, public toilets and waiting rooms in the American south during Jim Crow are historical reminders of obvious exclusion of a specific group of people (see Figure 4.2). These signs disappeared with the passage of federal civil rights legislation in the 1960s.

These examples point to easily recognizable signs/cues of exclusion, or inclusion, of specific groups of people. However, the design of a space can send perhaps less obvious, but nevertheless intentional, signals as to who is welcomed, who is expected to participate, and who will understand that the space was not meant for them. The tools that designers have at their disposal to create spaces are size, scale, shape, color, depth, arrangement, repetition, materials, and texture and sometimes placement, site, or location.

The ADA is meant to cover a wide range of disabilities (see https://adata.org/guide/americans-disabilities-act-questions-and-answers for

FIGURE E4.2 Signage at the bus station in Durham, North Carolina (1940).

Source: "At the bus station in Durham, North Carolina. 1940" (https://commons.wikimedia.org/wiki/File:At_the_bus_station_in_Durham,_North_Carolina._1940.jpg) by Jack Delano is marked as public domain under CC0 (https://commons.wikimedia.org/wiki/Template:PD-US), via Wikimedia Commons.)

a description of disabilities covered by the ADA); however, designed spaces often only accommodate those with physical, and visible or obvious, disabilities. The concept of universal design should be more expansive; but, again, designed spaces seem to concentrate on providing access to those with physical disabilities. The work of Magda Mostafa (2008) and Mostafa et al. (2023) provide a framework, ASPECTSS, for designers to create spaces for children and adults on the autism spectrum. Using the tools mentioned previously, the framework addresses acoustics, spatial sequencing, escape spaces, compartmentalization, transitions, sensory zoning, and safety. Magda proposes that designing within this framework creates spaces that ultimately provides those on the autism spectrum with the independence they need to reach their full potential. Social justice demands that everyone have this opportunity. Designers have a key role in creating this opportunity.

These are potentially powerful tools for designers, of all backgrounds, to use to respond to social justice issues. While every product or space is not meant to serve or be available to everyone (e.g., designing a child-friendly bar where alcohol is served is not a matter of social justice), intentional, or unintentional, design of a space or product that creates barriers for those who otherwise should be able to participate is a social justice issue.

The unequitable delivery/availability of healthcare in the United States is an example of a social justice imbalance. Where do people who live in marginalized communities get health care? What do the facilities look like? Do they invite people to enter, or do they further stigmatize people who need the services? A group in the San Francisco Bay area consisting of a local architect, healthcare providers, community activists, and others used design tools to create a public space for delivery of healthcare and other needed services to under-resourced communities of color. It is called the Pop-Up Village. The goal was not just to provide healthcare but to create spaces that encouraged the use of the services, was welcoming, and did not stigmatize those who sought the services. Their website states: "Place matters, and its power has been taken from those who stand to gain the most from its thoughtful creation. Rooted in the core values of Abundance, Action, Creativity, Generosity, Collaboration, and Participation, the Village seeks to bring regenerative design back into the hands of community" (https://popupvillage.org/about/).

Ensuring social justice in our society requires multifaceted approaches. The creation of place must be one of them. Every activity happens in a place (even if that place is digital/virtual), so the design of that place is critical to conveying who can participate in the activity. Those trained in the design professions have an abundance of tools at their disposal. One final tool necessary for social justice design is participation. Those who will be end users of designed places must be included in a meaningful way.

References

Chongkian. (2019). Nurul Muttaqin Mosque – Wheelchair ramp [Photograph]. *Wikimedia Commons*. https://commons.wikimedia.org/w/index.php?curid=75883311

Delano, J. (1940). At the bus station in Durham, North Carolina, 1940 [Photograph]. *Wikimedia Commons*. https://commons.wikimedia.org/wiki/File:At_the_bus_station_in_Durham,_North_Carolina._1940.jpg

Mostafa, M. (2008). An architecture for Autism: Concepts for design intervention for the autistic user. *International Journal of Architectural Research*, *2*(1), 189–211.

Mostafa, M., Sotelo, M., Honsberger, T., Honsberger, C., Lozott, E. B., & Shanok, N. (2023). The impact of ASPECTSS-based design intervention in autism school design: A case study. *International Journal of Architectural Research*. https://www.emerald.com/insight/content/doi/10.1108/ARCH-11-2022-0258/full/pdf?title=the-impact-of-aspectss-based-design-intervention-in-autism-school-design-a-case-study

HEALING ACROSS CULTURES

The Universality of Water

Connie Lin

The human attachment to water is universal. Humans have an affinity for landscapes containing water, or blue space. Simply by being near a water body, a sense of inner peace washes over us. Water exposure can be used as a strategy to reduce stress and anxiety, and therefore, most people associate water with a sense of inner peace. Regardless of one's cultural background or geographic location, all living beings understand the vitality associated with the presence of water. Alongside this need for survival, there is an innate human connection with water and its healing effects (Browning et al., 2014).

Research from regions with a large proportion of natural landscapes, such as in North America and Europe, has provided insights on the sensory impact and human interaction with physical water bodies (Triguero-Mas et al., 2015). Topics of discussion include accessibility, proximity, exposure, and interaction with blue spaces. It is suggested that humans are attracted to the visual rhythm of waves or ripples or the complex fractals in waterfalls (Török et al., 2015). Research from areas with larger urban densities, such as in Asia, provides insight on the implementation of artificial water features, digital displays, or sensory immersions (Xie et al., 2021). The consensus is that human interaction with artificial water features also provides positive effects for one's psychological and physiological health.

Although water has a global impact on health outcomes, what are the nuances of the effect of waterscape characteristics on human behavior? As a graduate student at Cornell University, I discovered that instead

DOI: 10.4324/9781003462019-10

of demographics such as nationality, race, gender, or age, it was more crucial to understand the type of waterscape, the duration of exposure, and the level of familiarity. I was pleasantly surprised when the results showed that both a slow-moving lake and a fast-paced waterfall were equally calming. While the footage displayed was a local water body in Upstate New York (see Figure E5.1), the participants dove into nostalgic thoughts about their own childhood experiences with water landscapes or safe spaces in various places. Participants shared their positive emotions, travel experiences, and peaceful spaces from around the globe with me, such as exploring the fjords in Norway, waves in Poland, waterfalls in Mexico, oceans in China, and lakes in India. After hearing about these experiences, there is something to be further explored about the increased empathy and sense of calm that are felt despite the cultural differences.

Although water can inspire inner peace, it is rarely used in high-anxiety situations such as healthcare facilities, due to sanitation and maintenance

FIGURE E5.1 Capturing display footage of Cayuga Lake, NY, USA.

issues. Bringing this research to a real-world design application, my research on the healing benefits of water was shared during the European Heritage Day in 2023 at the Centre Neuchâtelois de Psychiatrie in Préfargier, Switzerland (see Figure E5.2). Adjacent to the psychiatric hospital is a lake in which the proximity provides an easily accessible peaceful sanctuary for patients (Préfargier Site D'exception, 2023). The lake is described as a place of calmness, introspection, and self-reflection, as opposed to being a liability.

The affinity for water bodies such as oceans, lakes, waterfalls, rivers, and streams is a universal human trait, not a personal preference. When experiencing a new environment, it may feel uncomfortable and unfamiliar. However, the presence of water, ranging from a rippling lake or trickling fountain to a gushing waterfall or crashing ocean waves, can also bring peace.

FIGURE E5.2 Exhibition at the European Heritage Day 2023 posted at the Centre Neuchâtelois de Psychiatrie in Préfargier, Switzerland.

References

Browning, W., Ryan, C., & Clancy, J. (2014). *14 patterns of biophilic design*. Terrapin Bright Green.

Préfargier Site D'exception. (2023). *L'eau, des bienfaits même en version digitale ?* https://parcprefargier.ch/visite-didactique/leau-et-notre-sante/

Török, Á., Mestre, D., Honbolygó, F., Mallet, P., Pergandi, J.-M., & Csépe, V. (2015). It sounds real when you see it: Realistic sound source simulation in multimodal virtual environments. *Journal on Multimodal User Interfaces, 9*, 323–331. https://doi.org/10.1007/s12193-015-0185-4

Triguero-Mas, M., Dadvand, P., Cirach, M., Martínez, D., Medina, A., Mompart, A., Basagaña, X., Gražulevičienė, R., & Nieuwenhuijsen, M. J. (2015). Natural outdoor environments and mental and physical health: Relationships and mechanisms. *Environment International, 77*, 35–41. https://doi.org/10.1016/j.envint.2015.01.012

Xie, Q., Lee, C., Lu, Z., & Yuan, X. (2021). Interactions with artificial water features: A scoping review of health-related outcomes. *Landscape and Urban Planning, 215*, 104191. https://doi.org/10.1016/j.landurbplan.2021.104191

ARCHETYPES IN SACRED PLACES AS CONTRIBUTORS TO HEALTH AND WELL-BEING

Arsenio T. Rodrigues

Sacred places contribute significantly to our phenomenological experience of place on account of their inherent purpose and function and their relationship to peace. Among various place types, sacred places are speculated to significantly support healing (Alt, 2017; Perriam, 2015), help us feel energized (Brill, 1985), bring about a unique transformation of consciousness (Swan, 1990), and contribute to transcendental experiences (Critchlow, 1991). According to Brill (1985), specific emotions triggered and awakened at sacred places include ecstasy, ancient stirrings within the self, feelings of repose, feelings of sensory unification, and a sense of dissolution of self-emotions that are not only powerful but also intensely real, human, and similar to those experienced by archaic people in their sacred places.

Pioneering psychological studies have shown that transcendental experiences are characterized by a sense of ego surrender (James, 1902), feelings of unification and harmony with all things, feelings of blessedness or joy, a sense of timelessness and spacelessness, a sense of connecting with some sort of objectivity or ultimate reality, and a sense of divine presence or sacredness (Stace, 1960). Research involving neuroscience suggests that sacred experiences relate to elevated states of awareness and feelings of awe (Eberhard, 2005). According to Venolia (1988), such experiences, associated with feelings of calm, relaxation, and balance, are unique characteristics of healing environments. Our embodiment in sacred place, therefore, has the potential to contribute to our overall health and well-being and hence inner peace.

DOI: 10.4324/9781003462019-11

Place-archetypes are unique underlying characteristics that contribute to sacred place creation. Archetypes that contribute to sacred place creation could express higher intensions, exemplifications, and important cultural values. When these are uplifted in place, the resulting environment may be transformed from merely being secular to becoming sacred. Within this context, it is hypothesized that sacred places embody an exclusive set of place characteristics – a set of recurring archetypes that make them unique and by means of which they distinguish themselves from our other inventory of places. (Brill, 1985). Such sacred place-archetypes include:

- *Center* – focal point associated with ceremony, intense activity, and meaning.
- *Orientation* – significant building alignment with the cardinal directions.
- *Boundary* – enclosures that are substantial, clearly defined, physically fixed.
- *Descent* – features that signify a downward gesture with the earth.
- *Ascent* – features that express an upward gesture with the sky.
- *Void* – spatial voids that balance the buzz and detail along the periphery.
- *Numeric Order* – numerical identity, revealed as pattern in place.
- *Geometric Order* – shapes, embodying the transcendental root powers.
- *Spatial Order* – successive and rhythmic spatial organizations.
- *Anthropomorphic Order* – human-like attributes, proportions, and scale.
- *Celestial Order* – openings, markers that articulate the movement of the sun.
- *Flora* – landscape elements that are bordered and continually nurtured.
- *Materiality* – extraordinary materials that maintain their formal integrity.
- *Water* – pools, fountains, water bodies that evoke emotional responses.
- *Light* – illuminance that reveals form, contrasts with surrounding darkness.
- *Views* – openings that enable views between corresponding sacred places.
- *Passage* – thresholds, distinct transitional spaces between differing realms.
- *Ceremonial Order* – spaces that allow for ceremonies, meditation, rituals.

As I have previously discussed (Rodrigues, 2015), it is not merely the presence of these patterns but also the quality of their embodiment that significantly contributes to place being experienced as sacred. The famous rock garden at Ryoanji in Kyoto is acknowledged as a sacred place on account of its qualities that engender meditation. The rock garden displays several archetypes, including boundary, void, numeric order, geometric order, spatial order, flora, views, and ceremonial order (see Figure E6.1).

FIGURE E6.1 Rock Garden at Ryoanji, Kyoto.

The Salk Institute for Biological Studies (designed by Louis Kahn) in California might be classified as a secular place. However, a deeper and phenomenological investigation of the setting qualifies it as sacred place. The central plaza beautifully and seamlessly embodies center, orientation, boundary, void, numeric order, geometric order, spatial order, anthropomorphic order, celestial order, materiality, water, light, views, and ceremonial order (see Figure E6.2.)

Eliade (1957/1959) stated that it was around sacred place that archaic humans settled and founded their world, thereby giving meaning to sacred place as the center of the world. By living in close proximity to a sacred place and by merely entering it, we are assured of transcending the profane world, thereby sharing in the world of the sacred. In a time dominated by desacralization of place, the rediscovery of timeless archetypes and values, which have guided and informed place-making from time immemorial, is often overlooked or, worse, forgotten. These place-archetypes may therefore function as fundamental mnemonic triggers, which help us re-remember our quest for the most exemplary model of place – place that is sacred and place that heals – places of peace.

FIGURE E6.2 Central plaza at Salk Institute for Biological Studies, CA.

References

Alt, P. L. (2017). Sacred space and the healing journey. *Annals Palliative Medicine, 6*(3), 284–296. https://doi.org/10.21037/apm.2017.06.09

Brill, M. (1985, September 20). *Using the place-creation myth to develop design guidelines for sacred space [Conference address]*. Annual Conference of the Council of Educators in Landscape Architecture at the University of Illinois, Urbana, IL.

Critchlow, K. (1991). Space: Plenum or abyss. In K. Vatsyayan (Ed.), *Concepts of space, ancient and modern* (pp.149–156). Abhinav Publications.

Eberhard, J. (2005). Sacred spaces: A neuroscience perspective of sacred spaces. *American Institute of Architecture Journal, 1*(5), 6–7.

Eliade, M. (1959). *The sacred and the profane: The nature of religion* (W. R. Trask, Trans.). Harcourt, Inc. (Original work published 1957)

James, W. (1902). *Varieties of religious experiences*. Modern Library.

Perriam, G. (2015). Sacred spaces, healing places: Therapeutic landscapes of spiritual significance. *Journal of Medical Humanities, 36*, 19–33. https://doi.org/10.1007/s10912-014-9318-0

Rodrigues, A. (2015). Sacred place-making: The presence and quality of archetypal design principles in sacred place. In D. Galbreath (Ed.), *The changing world religion map: Sacred places, identities, practices and politics* (pp. 2377–2398). Springer. https://doi.org/10.1007/978-94-017-9376-6_125

Stace, W. (1960). *Mysticism and philosophy*. Lippincott.

Swan, J. (1990). *Sacred places: How the living earth seeks our friendship*. Bear and Company, Inc.

Venolia, C. (1988). *Healing environments: Your guide to indoor well-being*. Celestial Arts.

ENVIRONMENT AS A CONTRIBUTOR TO RECOVERY AND PEACE

Marie Elf

The search for inner peace is a journey that almost every person embarks on at some point in life. For individuals who have experienced a stroke, this endeavor takes on new meaning, not only in the mind but also in the physical space. Virginia Woolf once spoke of the significance of "a room of one's own." For her, this was a space for creativity and being able to express herself in text. For those recovering from a stroke, a supportive home, room, or environment can be a place for healing, self-discovery, and reconnection with society. The person with a stroke is not just a passive recipient in this environment but an active participant who uses, makes choices, and makes decisions about the space and daily life. The environment of the home does more than meet physical needs. It offers a place of inner peace, a space to reconnect with one's identity beyond the label "stroke survivor."

People with a stroke often struggle with cognitive, physical, and emotional issues that can result in a changed relationship with their surroundings. A once-familiar home can become a challenging place requiring new navigation. It can be about doors that used to be easy to open but now appear as barriers; stairs can feel like mountains, and even reaching for a favorite book on a high shelf becomes challenging.

In my research in Sweden. I have met numerous individuals struggling to adapt to their homes and reintegrate into society post-stroke. One woman was unable to use the bus due to its complex environment, with stairs and paying system. She resorted to not using the bus,

DOI: 10.4324/9781003462019-12

FIGURE E7.1 Simple interactions, such as traveling by bus, can be challenging for those who have experienced a stroke.

with her need to go to other places restrained. Similarly, I met a man who had carefully nurtured his garden throughout his adult life but could not reach it due to accessibility issues (Kylén et al., 2022) (see Figure 7.1).

These personal stories underscore the pressing need for environments to be integral to rehabilitation. If society recognized the significance of supportive environments as a component of recovery, rehabilitation could adopt a more holistic approach, bridging the gap between physical healing and meaningful life recovery.

However, among the challenges that individuals with stroke might face, the environment also presents opportunities for recovery. A space tailored to one's needs can transform from merely a room into a place of comfort and independence. Such a space can also serve as a familiar and natural setting for rehabilitation. Everyday objects and personal devices can be used as therapeutic tools and can be just as effective as devices in institutional settings. Simple modifications of the home can lead to significant improvements. For instance, ramps can replace stairs to ensure accessibility; lowered shelves can make everyday items readily accessible, fostering autonomy; gentle, peaceful colors and textures can envelop the senses, inducing a sense of calm (Connor et al., 2023).

In essence, the environment, when consciously adapted, has the potential to be a powerful link in a person's recovery. It is not just about having your room but having a room you understand and that nurtures and strengthens your identity – a space that does not mean the end of a journey but a continuous healing process. At home, there may be an opportunity to rediscover interests, whether reading, painting, or simply watching the world go by from a window (Figure 7.2).

While adapting individual spaces is essential, it is equally crucial for society to design inclusive environments. The journey of recovery is not just an individual's endeavor but a collective responsibility, fostering connections and ensuring every room, every space, is one where individuals can heal, rediscover, and thrive.

FIGURE E7.2 Minor modifications to the home environment can result in improved experiences.

References

Connor, E. O., Dolan, E., Horgan, F., Galvin, R., & Robinson, K. (2023). A qualitative evidence synthesis exploring people after stroke, family members, careers and healthcare professionals' experiences of early supported discharge (ESD) after stroke. *PLoS One*, *18*(2), e0281583. https://doi.org/10.1371/journal.pone.0281583

Kylén, M., Ytterberg, C., von Koch, L., & Elf, M. (2022). How is the environment integrated into post-stroke rehabilitation? A qualitative study among community-dwelling persons with stroke who receive home rehabilitation in Sweden. *Health & Social Care in the Community*, *30*(5), 1933–1943. https://doi.org/10.1371/journal.pone.0281583

SEXUAL HEALTH ENVIRONMENTS AND PRIVACY

Evangelia Chrysikou

Good health is critical to personal peace, and sexual health (SH) is key to physical and mental health. Unfortunately, the UK has a high rate of late diagnosis of HIV (NHS England, 2022). Stigma or fear prohibits people from being tested, as people might be concerned about lack of personal privacy and the impact of the results on others.

Fear, humiliation and disempowerment, even brought on by doctors, can be significant barriers to accessing healthcare (Malterud, 2005) and a disruption to personal peace. The stigma associated with SH amplifies this barrier, which is further exacerbated by medical buildings that reproduce mechanisms of power and control (Markus, 1993). Eliminating privacy, implemented in asylums incarcerating people with syphilis such as Panopticon (see Godbey, 2000), has been one of those. Today, privacy is fundamental for SH provision (Heath et al., 2023; Warr & Hillier, 2008), although not always practiced. Understanding how it can be designed and implemented is essential for patients' journey to recovery and personal peace (see Figure E8.1).

Once I visited a sexual health ward in London, being in the process for in vitro fertilization (IVF). Upon arrival, I completed a form requiring an exhaustive account of my health and sexual life. I had to grab a chair from those forming a sociofugal link around the walls. I covered my replies to protect my territoriality. I longed for some positive distraction. I ended up observing people, invading their personal space bubble. People's forms were in my

DOI: 10.4324/9781003462019-13

FIGURE E8.1 Counter design that enables patients in a sexual health clinic to enter their medical information privately.

visual field. I fixed my gaze at my lap. To hear each other, an employee and a visitor, separated by clear plastic screens, spoke loudly:

(employee):	Do you know why you are here?
(girl):	No.
(employee):	X tested positive for HIV. He said you had sex with him . . .

HIV sounds like death sentence. Being told in public is devastating (Malterud & Hollnagel, 2007). We became unwilling participants in the scene. I too explained why I was there and moved to another waiting area. The female area was full. I was sent to the male. It was bare, artificially lit, without windows. Later, I was escorted to a tiny examination room, with charts and photos of infected body parts. A droplet's distance from the examination bed, a urinal self-flushed every 15 minutes. Droplets invaded my space.

Alternatively, a woman might seek maternity wards for SH testing. Some women attempt IVF after having a miscarriage but are reluctant to pursue this option due to the co-location with women experiencing successful pregnancies. Women from one culture might not feel comfortable in certain health services architecturally designed to encourage another culture (Chiotti & Joseph, 1995). Contemporary aesthetics may encourage some people and discourage others. SH facilities need to accommodate victims of trauma, such as those who have experienced sexual abuse or the impacts of violence. War immigrants and refugees are disproportionately affected by Sexually Transmitted Diseases (STGs) (McGinn, 2000). A sense of privacy and control is indispensable. We all need a place to connect with our inner self to achieve wellbeing and inner peace and allow those who care for us in, on our terms.

As SH is one of the most stigmatized areas of health, the associated facilities must incorporate psychosocially supportive features inside generic services that convey neutrality. This specialized design prototype constitutes an area in which design could aid health equity and justice. Space and society are deeply interlinked. Supporting people's sexual health, we foster inner peace.

References

Chiotti, Q. P., & Joseph A. E. (1995). Casey House: Interpreting the location of a Toronto AIDS hospice. *Social Sciences Medicine, 41*(1), 131–140. https://doi.org/10.1016/0277-9536(94)00304-c

Godbey, E. (2000). Picture me sane: Photography and the magic lantern in a nineteenth-century asylum. *American Studies, 41*(1), 31–69.

Heath, G., Clarke, R., Ross, J., & Farrow, C. (2023). Factors influencing non-attendance at sexual healthcare appointments in the UK: A qualitative study. *Sex Health, 20*(5), 461–469. https://doi.org/10.1071/sh23099

Malterud, K. (2005). Humiliation instead of care? *Lancet, 366*(9488), 785–786. https://doi.org/10.1016/s0140-6736(05)67031-6

Malterud, K., & Hollnagel, H. (2007). Avoiding humiliations in the clinical encounter. *Scandinavian Journal of Primary Health Care, 25*(2), 69–74. https://doi.org/10.1080/02813430701237721

Markus, T. (1993). *Buildings and power: Freedom and control in the origin of modern building types*. Routledge.

McGinn, T. (2000). Reproductive health of war-affected populations: What do we know? *International Family Planning Perspectives, 26*(4), 174–180. https://doi.org/10.2307/2648255

NHS England. (2022). *NHS HIV testing rollout identifies hundreds of new cases*. https://www.england.nhs.uk/2022/11/nhs-hiv-testing-rollout-identifies-hundreds-of-new-cases/.

Warr, D., & Hillier, L. (2008). That's the problem with living in a small town': Privacy and sexual health issues for young rural people. *The Australian Journal of Rural Health, 5*(3), 132–139. https://doi.org/10.1111/j.1440-1584.1997.tb00254.x

ARTICLES OF DISPLACEMENT

Denise Nicole Green

Creating conditions of possibility for peace through design involves embracing, seeking, and making space for seemingly divergent forces. In a world governed by divisive thinking, taking a "both/and" approach when facing false dichotomies is the first step towards equanimity and equilibrium. Holding space for "both/and" tensions often involves rupture and subsequent transformation (Smith & Lewis, 2022). The synergies produced by dynamic tensions unravel binaries and offer new design possibilities. In this essay, I discuss the tensions and contradictions that fueled my creative spirit and understanding of peace through a naturally-dyed fashion collection, *Articles of Displacement*, which I co-created with plants, people, and places in 2022 (see Figure E9.1).

The fashion industry is broken and the source of immense harm to people and planet (Rovira & Domingo, 2019; Tounsadi et al., 2020; Yaseen & Scholz, 2019). At the same time, fashion can bring joy into our daily lives, be a driver of social change, raise awareness of injustices, and empower individuals and communities. Fashion itself is both good and evil and fueled by ambivalence (Kaiser & Green, 2022). My creative and scholarly design practice centers these contradictions and complexities, and I began the design process with questioning: What is the antidote to colonialism and settler mentalities? Unsettling. What is the antidote of anger and hatred? Joy. How can we create peace through disruption? By staying committed to delight as the driving force behind design (see Figures E9.2, E9.3, and E9.4).

I took these answers to heart by de-centering myself as designer and bringing the hopes and dreams of the wearers and materials to the fore.

DOI: 10.4324/9781003462019-14

FIGURE E9.1 Designing with Ahava Irie Mae Sapp (age 6) in the summer of 2022.

Sources: Jahnna Edelman. Photo of Ahava Irie Mae Sapp and Denise Nicole Green by Jahnna Edelman.

I challenged approaches to fashion that have tended to prioritize the prosperity of corporations and the egos of designers and instead wondered how to co-create through elevation of all participants in the process. As a result, *Articles of Displacement* is both celebratory and truculent. More questions soon followed and guided my process: How will relationships to plants, place, family, community, and nation shift as populations are forced into motion by climate change? What becomes of

FIGURE E9.2 Ahava Irie Mae Sapp models the garment she helped to co-design in December 2022. Dress is a dyed with avocado pits and madder root and contact-dyed with coreopsis and purple basil.

Source: Photo by Simon Wheeler for Cornell University.

FIGURE E9.3 Embodied expressions of joy through fashion. The jacket is indigo-dyed silk charmeuse with batik wax resist. The leggings and tunic dress are dyed with turmeric and contact-dyed with yellow onion skins.

Source: Photo by Simon Wheeler for Cornell University.

FIGURE E9.4 Sharing in style and happiness: leggings are nylon-spandex blends vat-dyed with logwood and contact-dyed with red onion skins and dahlias. The sweaters are nylon fleece and dyed with logwood and madder root.

Source: Photo by Simon Wheeler for Cornell University.

local communities; ideas about naturalization and Indigeneity; social and cultural identities; and, at a fundamental level, connections to landscapes, atmospheric conditions, seasonal cycles, and shared environments with flora and fauna? How might clothing design be guided by rapidly changing social, physical, and aesthetic ecologies? The answers, this time, came by facing anthropocentrism and empowering materials. Decentering the individual human (both "designer/producer" and "wearer/consumer") opened space for a new approach to design that was instead guided by shared ecologies (i.e., the plants, places, and communities around us and part of us). By displacing traditional approaches to design that tend to prioritize humans, we empowered non-human actors, ecologies, and materials in a collaborative, relational, and responsive design process (Latour, 2007).

Through responsive design, the collection sought to challenge some of the colonial and capitalist regimes that fueled fashion's role in producing the global climate crisis. Those systems include industrialization of fiber production through large-scale agriculture, corporate chemical engineering, synthetic dyeing and printing processes, designed and artificial obsolescence of clothing, and the exploitation of labor and environments in the mass-manufacture of apparel. The collection also honors the plants, people, and places that made co-creation possible. Peace came through human and non-human actors coming together through delight, collectivism, and self-determination. All garments were naturally dyed with bio-based, renewable sources, including those from food waste, invasive species, plant pests, local fauna, and plants grown in small gardens.

Articles of Displacement centers exuberance as an articulation point between social and climate justice. The designs grapple with both structural challenges and individual agency. The availability of dyestuffs, the size, shape, and eventual coloration of each textile, anticipated environmental conditions and bodily movements, the needs and desires of the wearer, and commitment to material longevity were just some of the factors shaping the design process. De-centering and displacing profit motivation and anthropocentrism enabled shared ecologies (i.e., the habitats, natural resources, and communities around us and part of us) to guide the design and production process. For example, the clothes were created in response to the fabric's surface design, which was often hard to predict since contact-dyeing techniques can create unexpected results. Embracing the joy of spontaneity and the unknown brought peace into the process of relinquishing control.

While natural dyes used in the collection were renewable, the fabrics told a more complex and paradoxical story. Nearly all garments in the collection were made from silk or nylon. Considered "twin polyamides," the two fibers share similar chemistry but have a dramatically different impact on the environment. Petroleum-based nylon was the first commercially successful thermoplastic polymer and shaped the landscape of industrial manufacturing in the mid-20th century. Blending nylon with spandex accelerated the growth of activewear apparel since the 1970s. Unlike silk, which is easily weakened by ultraviolet light, heat, and abrasion, nylon is one of the most durable, lasting fibers. By using nylon – that is, one of the culprits of climate change – the designs address industrial capitalism's penumbra and the products innovated under these regimes.

Coming to peace means recognizing the murky nature of sustainable fashion production and acknowledging that nothing made of new materials can actually be sustainable. But we can move towards greater sustainability by designing for delight, which in turn means designing for longevity and joyful usability. Design can also (re)build relationships with the natural world and raise awareness of ecological pitfalls. By connecting to plants and other raw materials required for dyeing, creating inter-reliant systems of production through co-design, and employing some of the products of the environmental plight, *Articles of Displacement* imagines a future where design is a collective, material, embodied, inter-species, place-making, and community-creating endeavor that produces peace.

References

Kaiser, S. B., & Green, D. N. (2022). *Fashion and cultural studies* (2nd ed.). Bloomsbury Publishing.

Latour, B. (2007). *Reassembling the social: An introduction to actor-network-theory*. Oxford University Press.

Rovira, J., & Domingo, J. L. (2019). Human health risks due to exposure to inorganic and organic chemicals from textiles: A review. *Environmental Research*, *168*, 62–69. http://dx.doi.org/10.1016/j.envres.2018.09.027

Smith, W. K., & Lewis, M. W. (2022). *Both/And thinking: Embracing creative tensions to solve your toughest problems*. Harvard Business Review Press.

Tounsadi, H., Metarfi, Y., Taleb, M., El Rhazi, K., & Rais, Z. (2020). Impact of chemical substances used in textile industry on the employee's health. *Epidemiological Study Ecotoxicology and Environmental Safety*, *197*, 110594. https://doi.org/10.1016/j.ecoenv.2020.110594

Yaseen, D. A., & Scholz, M. (2019). Textile dye wastewater characteristics and constituents of synthetic effluents: A critical review. *International Journal of Environmental Science and Technology*, *16*, 1193–1226. https://doi.org/10.1007/s13762-018-2130-z

5

PEACE FOR A LIFE INDOORS

Rosalyn Cama

Imagine walking through a doorway and immediately being struck by a feeling, and then realizing that something about the environment is fostering your wellbeing. Feeling is not just tactile – it is sensing; it is a neurological state that engulfs you, potentially distributing positive feelings of joy or peacefulness, or simply being alive. In Gawande's words,

> We have been wrong about what our job is in medicine. We think our job is to ensure health and survival. But really it is larger than that. It is to enable wellbeing. And wellbeing is about the reason one wishes to be alive.
>
> *(2014, p. 259)*

What is it that nudges that feeling of mental wellbeing that is peacefulness, and cues you to know that you will navigate and occupy a space peacefully with others? This question was the basis of my quest for peace for life indoors.

Peace

'Peace' has come to be an endearing greeting with two interpretations. If its sign is made with one's head bowed and hands held together at one's heart, it is meant as a wish for inner calm or personal wellbeing. However, if it is made with two forefingers held up in a V, it is meant as a wish to exist collectively, in harmony.

DOI: 10.4324/9781003462019-15

FIGURE 5.1 Interior design detailing that uses "naturalistic tendencies" (Wilson & Kellert, 1993, p. 45) nudging spiritual peace. Temple Beth Tikvah in Madison, CT (Interior Design: CAMA, Inc., Artist: Linda Cummings, Architect: Duo Dickinson).

Source: "GENESIS" Linda Cummings © 2024. Ark Door, Temple Beth Tikvah (H). Trans-illuminated Art Glass 120" × 84"

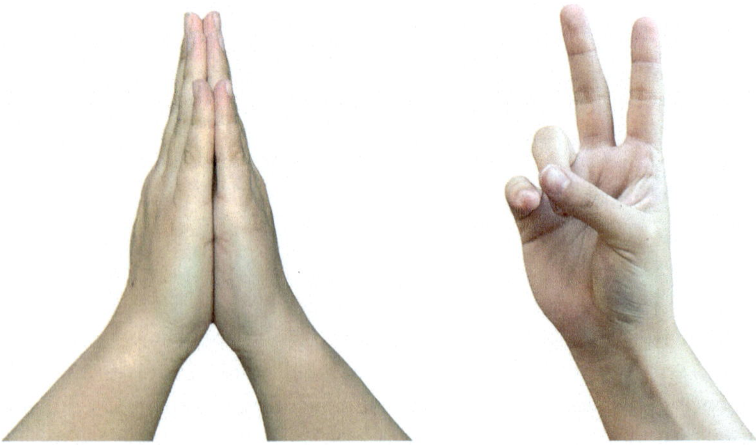

FIGURE 5.2 Gestures for peace. Peace as a verb extending wishes for personal well-being (left) and peace as a noun extending wishes for communal harmony (right).

Based on its root word, peace, peacefulness in the design of a built environment is a noun that indicates harmony in personal relations (Merriam-Webster, 2023). There are design principles that are known to nudge social behaviors like decision-making while shopping (e.g., Naylor, 2008). If people could be nudged to be peaceful as a result of changes in the design of their environments, then interior designers could be charged with achieving personal wellbeing by way of communal harmony.

Sartwell (2004) points us toward several aspects of communal harmony in his definition of ultimate values:

> The classic triad of ultimate values is truth, goodness, and beauty. Truth into belief: to believe something is to take it to be true. Goodness is built into decision: to decide on a course of action is to decide that it is best. And beauty is built into desire: what we desire we learn to find beautiful.
>
> *(p. 5)*

If the elements of Sartwell's (2004) triad could be examined individually and applied to life indoors, then one could say that truth can be clearly stated in every building project's vision. Those stated values or beliefs should be held throughout all decision-making steps, even at the ends of projects. Decisions should never fully rest on cost or value engineering. And just as safety has been at the top of all healthcare projects' lists for the last 20 years, might not peace for mental wellbeing now move itself to the building decision-makers' top-of-mind vison guidelines? One could argue that goodness comes from basing design decisions on client's beliefs measured through a baseline of peer-reviewed evidence. And last, acknowledging beauty, quite simply, as that which attracts us or nudges behaviors as seen through the lens of truth and goodness. The values of truth, goodness, and beauty all contribute to mental health, wellness, and peace in the lives of people indoors.

If we are striving for an outcome of personal wellbeing in communal harmony when in an interior environment, then we need to fully investigate how that comes about, how it can be aligned with basic interior design principles, and how interior designers can achieve that harmony.

Truth Provides Vision

The most important step in the design process is setting measurable goals aligned with a values-driven vision for a project. In the field of healthcare, where innovation is often triggered by a public health need, there are many examples of how measurable goals can be set and lead to new visions of how peace in interior design can be achieved. One such example is the Pebble Project, a research project begun by the Center for Health Design (n.d.). The purpose of the project was to generate a growing body of evidence regarding innovative design interventions that would create a "ripple effect," so that one investigation would beget the next and trigger the refinement of the design principles, spawning customized designs. It was thought that when followed by an iterative process amongst academic researchers and multi-disciplinary design professionals, advancements in interior design principles would occur. New trends would be published, noticed, and adopted by other practitioners. When the benefits of these trends became truly evident through business case studies, then these design principles could be accepted, scrutinized, and written into code. The next innovative design team then would have a springboard set at a higher level to explore what comes next. The process would repeat itself again and again, and advancements would be made. The Pebble Project is less active now, but, when multiple teams engage in this fashion, their shared intelligence raises the bar for the next cohort of values-driven teams.

By raising the bar from understanding only current practices to developing an awareness of the baseline findings of new developments in the field, then, and only then, can designers improve the human condition through the design of the built environment. This design methodology is already in place in the healthcare field, offering growth and feedback for deeper research and design acceptability. The continuous loop of design research and innovative design thinking is the way to advance the development of interior environments that promote mental health and peacefulness.

One dilemma is that, as a shift in design thinking occurs, then project managers must work to overcome the perception that non-conventional design principles are too expensive or unattainable in readily available product offerings. To reach the tipping point, where managers can see the value in new approaches that support peacefulness, designers must show a solid return on its investment. The clues are available, but the trail is new. The profession needs a catalyst.

Vision for a project can come from many directions and involve the wishes of many people who interact with a space. Asking clients what is needed for a new design project often yields suggestions that may replicate a dysfunctional current scheme. Instead, ask clients to dream and tell a story that aligns their vision with an enriched transformational interior experience. Have them look to the edges of society for examples of baseline acceptance to elevate their intellect. Storytelling gives people permission to look at their world differently and challenge their assumptions. It is in their storytelling that design teams can plan scenarios and become imagineers that open new possibilities. The questions formulated in their speculations fuel future research questions and give new ideas momentum. If proven, dreams, once part of a fantasy, are given permission to become mainstream (Wacker et al., 2000; Taylor & Wacker, 1997). Where comfort is created, permission for innovation will follow. An innovative solution can address a client's beliefs, not simply be the result of a designer's unfounded creative whim. Clearly stating the truth in a project advances its impact on those who will enter the building and influences projects that will follow.

Goodness and Beauty

In healthcare, several well-studied projects, such as the Pebble Projects, provide evidence to support the view that "form follows function" in great design. Studying the less functional and more experiential transformations of human experiences, as in wellbeing and harmony, requires more complex research projects. These types of projects need to simultaneously explore the physical, neurological, and spiritual responses one has to the built environment. The goal of a designer is not merely to have an occupant survive in this kind of interior experience but rather, to Dr. Gawande's point, to enjoy it. This body of knowledge needs more depth to influence decision-makers and practitioners.

Beauty has been known to transform the human experience. To achieve peace for life indoors, one must begin by identifying existing environments that are thought to be beautiful so that they might be studied to build a baseline of knowledge and scrutinized to see where further research is needed. One often thinks of museums, luxury properties, and religious temples, such as the one in Figure 5.1, as examples of beauty that elicit transformative feelings.

FIGURE 5.3 Transformational experience using beauty through goodness. Beinecke Rare Books Library at Yale University in New Haven, Connecticut (Architect: Gordon Bunshaft, FAIA, SOM).

Source: Photo by Iwan Baan.

They typically include complex and expensive design details. One such interior environment that comes to mind is the Beinecke Rare Books Library in New Haven, Connecticut. It was funded by a benefactor for the Yale University campus and has an interior environment that was designed to protect rare manuscripts from decay by filtering out ultraviolet light using translucent Vermont marble as an evidence-based design intervention.

The manuscripts are encased in environmentally secured glass stacks, which are protected by this exterior skin of translucent marble, causing those who enter through the library doors to experience a magical feeling of filtered sunlit illumination. The sensitivity of transformational design thinking, influenced by his research, guided Gordon Bunshaft, FAIA, not only to preserve the manuscripts but also to dramatize their discovery. The choice to use sunlight through a natural filter delivered the desired functional outcome (preserving the encased manuscripts from physical decay – truth). However, this naturalistic design solution also elevated the building occupants' experience in the reverent study of manuscripts by co-existing in an equal state of peace and harmony in the space – beauty through goodness.

This well-designed interior is full of beauty, and beauty changes lives. Bunshaft understood that lighting can be more than the typical lamping specified throughout a facility. He convinced owners that to align with one's ultimate vision and values, one must take a risk.

My Story

To demonstrate an awareness of how nature's dynamic diversity in an environment can contribute to feelings of peace at a particularly difficult time, I want to tell my story. It is a story about the omni-present feeling of grief in surviving the loss of a spouse and finding peacefulness in a special environment.

The setting is in an iconic piece of architecture designed by my mentor, Vincent Amore, AIA, located at the edge of Long Island Sound, facing the southwest prevailing breezes of the Connecticut shoreline. The experience, inside and out, is that of a dry-docked boat. I called it my Peace House – it is a peaceful place connected to cherished memories.

My husband Ron was a competitive sailor and knew the waters of Long Island Sound, its currents, and the prevailing pattern of wind shifts. In the autumn after he died, I would go to my proclaimed "peaceful place" connecting our cherished memories of our time sailing together. When the season changed, and I came to spend more time inside, I would sit and observe the patterns of light caused by the sun and moon rising and setting on the water. The days changed to months and the seasons changed four times, yet as the patterns of sun, wind, and water changed, my sense of escape from grief remained the same – peaceful.

FIGURE 5.4 Peace House on Grass Island, Guilford, CT (Architect: Vincent Amore, AIA).

Source: Dennis M. Carbo Photography.

During the first seasonal change, I was invited to write this chapter. It was an ironic request, as I was seeking my own peace in a new life pattern. Truth be known, I hated being alone, so I invited everyone to experience this slice of Ron's harmony in nature with me; it was a place that offered joy when grief prevailed. My visitors and I were enriched in this place, where the inside/outside lines became blurred. I understood that my visitors were a distraction, but what role did the place play? I noticed that I was not just surviving grief but, oddly, finding enjoyment whether alone or with others at the edge of the sea.

I asked myself, why was this a peaceful place? Was it the iconic design? Was it its position in nature? I read Gooley's (2017) book on *How to Read Nature* and then would sit silently alone to understand what was happening in this place. While trying one of Gooley's exercises, I was struck by something I had discovered in my firm's work on the design of the IOA Cama Glider, a therapeutic chair that slides forward and back. I discovered there truly is joy in movement, about which I knew little, except for its ability to provide a positive distraction.

This is the Gooley exercise:

> While indoors, close your eyes and breathe ten times, deeply and slowly, in through your nostrils and out through your nostrils . . . Now repeat the

> exercise outdoors. Ten breaths, deeply, slowly, nostrils only. . . . Compare the experiences.
>
> *(p. 26)*

It was while performing this exercise that I experienced nature's dynamic movement. It hit me that there was more to the long view of nature, where I found solace. It was nature's complex patterns of motion that held my attention as a peaceful form of positive distraction.

The experience I discovered was found in the southwest prevailing breeze as it waxed and waned, bringing salt air moisture into my breath. The sun warmed what a northerly breeze cooled. The sound of the rustling sea grasses laid down the melody for the sporadic bird song lyrics. The cadence of the waves lapping against the stones set a rhythm to wind down my heartbeat. What really mesmerized me and bonded my soul in a state of peace was the glint of the sea. I first noticed it on a cold winter's day as the sun moved in and out of clouds. The vibrant white light reflected on the sea, and, as the clouds moved to let the sun's rays through, the sea's joyful glint rushed towards me with the sparkle of diamonds. It hit me that the glint only came when the wind moved across the water to reflect and refract the sun's light peeking through the clouds. I realized that it was movement – movement created through the dynamic components of nature, all working in unison, like a well-conducted symphony. It was nature's symphony that brought me peace!

Based on that experience, I pivoted my thinking about finding peace for a life indoors. I realized that it could not be found in the transactional stagnancy of our built environment's inanimate objects, but it can be found in the diversity and the symphony of nature's living palette. It was in the palette of Vermont marble's magical interaction with the cycle of the sun, only augmented by electric light as a background intervention when the sun disappeared. Interior design palettes need to work in unison with nature to transform our experiences and allow our mind, body, and spirit to become alive.

In lectures over the last 25 years, 95% of my interior design firm, CAMA, Incorporated (CAMA), audiences chose an outdoor environment rather than an interior environment to relieve stress and promote wellbeing. It's a fun fact that many in the audiences are architects and interior designers! Ironic – yes, especially since many are evidence-based healthcare professionals. The biophilia hypothesis states that humans have "a genetic bias for this naturalistic tendency" (Wilson & Kellert, 1993, p. 45). "The mental benefits of these activities (naturalistic experiences) have been related to tension in the release for relaxation, peace of mind, and enhanced creativity derived from the observation of diversity in nature" (p. 46). Diversity in sensory engagement is therefore a major clue for interior designers who wish for their designs to trigger these same responses. Does nature need to be present or replicated, or can a designer merely understand the properties that trigger a similar response and

FIGURE 5.5 Glint, found in the diversity and the symphony of nature's palette.

Source: Image taken in a Long Island estuary. "JUNE BUG VII" Linda Cummings © 2010. Archival Pigment Print.

recreate them with engineered systems? If we are genetically predisposed to be outdoors, then what clues are there that lead to a better understanding of the subtleties, not just a recreated long view of these naturalistic experiences indoors? In Richard Prum's book *The Evolution of Beauty* (2018), he states, "To be a more rigorous discipline, aesthetic philosophy must grapple with the full complexity of the biological world" (p. 555). It is complicated.

An Interior Designer's Changing Role

Why nature's dynamic motion soothes us is beyond the practicing interior designer's capability to answer. Is the human body more comfortable in a vibrant environment rather than a stagnant, climate-controlled interior? Kellert's *Biophilic Design* (in Heerwagen & Gregory, 2008) states that psychiatrist Aaron

Katcher refers to motion perceived as "Heraclitean movement," describing it as "always changing and remains the same" (p. 231). It signals comfort and safety. Was that what I was experiencing – safety in a place my husband and I shared and hence in which I found escape or comfort and subsequently peace? Is this a neurological or psychological response programed into our evolutionary DNA, that is, as Wilson suggests, the key to a spiritual connection with nature? Did I feel that in this place I found a memory where I had not lost the essence of my sailor? Is it too much to ask a researcher if the energy found in nature speaks to one's energy known as spirit? A rabbi I worked with on the design of Temple Beth Tikvah told me about a reference written in the Bible that uses the Hebrew word "Ruach," referring to God as a breath, a wind, or a life force that sustains all living things, human beings included (Bolinger, 2020). Sustains? Brings joy? How many times have you rushed to see a beautiful rising or setting sun? Biophilic design (design that connects one with nature) lays the groundwork for a human response to nature's diverse patterns and interactive motions. Can interior designers take this thread of knowledge and use it as a springboard for incentivizing the next level of questions to be answered by our academic colleagues and their students? Developing biophilic interior design principles with standard interior design interventions capable of replicating these enjoyable qualities is now a personal quest for me. As the solutions are multi-disciplinary, I begin to see an interior designer's role shifting to one of conductor – the conductor of the environmental symphony made up of a building's sensory systems and material applications to deliver a peaceful experience.

Adding peace to our health, safety, and welfare codes requires design thinking, focused research, multi-disciplinary design skill, and salesmanship. The paradigm shift will occur when codes, like Gawande's wish for the field of medicine, graduate from just ensuring survival to requiring enjoyment in our life indoors where we spend most of our time being alive. Our DNA has been primed for survival outdoors and must not be lost in the devolution of our current building practices. Human life safety codes can be expanded to include being alive indoors in wellness and peace.

What if we started with how to design interiors so that the human mind and spirit trigger personal enjoyment in addition to physical safety? Might a different body of knowledge emerge? This shift needs a timely cause. This unmet need for peace for a life indoors requires context, and action requires that the timing of change be right. Technology and the control systems exist to replicate all modulating qualities in a preferred environment, but they are rarely programmed for such. The largest hurdle may be the economic stressors on a project. Budgets and schedules dumb down poorly stated visionary programs into over-programmed minimally required code solutions. Currently costs are prohibitive, perhaps because there is too much vested into the traditional systems of building to expect change for the sake of delivering peace unless the timing is right. Truth, goodness, and beauty are forgotten.

Post-pandemic human behaviors provide an excellent example of a cause whose timing triggered the search for a new paradigm. The United States Surgeon General, Dr. Vivek Murthy, released a Surgeon General Advisory calling attention to the public health crisis of loneliness, isolation, and lack of connection in our country. Even before the onset of the COVID-19 pandemic, approximately half of US adults reported experiencing measurable levels of loneliness (US Department of Health & Human Services, May 3, 2023). Society is currently ripe for a tipping point. It is clamoring for wellbeing and harmony, an affirmation of a potential change. Malcolm Gladwell states that "Tipping points are the reaffirmation of the potential change and the power of intelligent action" (2000, p. 259). Society is clamoring for wellbeing and harmony – an affirmation of a potential change. This opportunity for intelligent action to overcome the lack of peacefulness in occupants in code-compliant environments can come from a small few. Yes, those interested enough to read this chapter are being summoned. The call is to listen for the unmet need, question it, and ask what is really happening in the design of the built environment that can nudge the human body, mind, and spirit ergonomically, neurologically, and psychologically.

The wellbeing of individuals, as well as their harmonious interpersonal connections with others, is important, as most people today spend the great majority of their time indoors. Current interior design trends are heavily weighted toward physical wellness and environmental sustainability. The evidence-based design movement has expanded and integrated the responsibility of the professional interior designer by allowing multi-disciplinary teams to co-design.

Multidisciplinary teams can include engineers, lighting designers, acousticians, architects, landscape architects, interior designers, code consultants, fund developers, and many others in each specialty area of design. However, the most intentional transformative design principles often succumb to the minimum requirements posed by life safety codes. What if the life safety codes added requirements that addressed peace for life indoors?

One could argue that peace is a life safety issue yet to be stated as such in building codes. According to the World Health Organization (WHO), health and peace are interrelated. The director general of WHO, Dr Tedros, states, "there cannot be health without peace, and there cannot be peace without health" (WHO, n.d., p. 1). The World Health Organization constitution states: "Health is a state of complete physical, mental and social well-being and not merely the absence of disease or infirmity" (WHO, 2023, p. 1). An important implication of this definition is that mental health is more than just the absence of mental disorders or disabilities.

A design collaborative should align with the WHO by accepting peace as a factor in health. This kind of alignment would give our desired outcome the reverence it needs, much like what was done in 1999 for safety when groups that demanded that safety was the primary issue for healthcare design.

Their demands were well timed, were aligned with the Institute of Medicine's (IOM) (2001) aims, and fueled the agenda of many in charge of the billions of dollars of construction in hospital building at the completion of the second millennium.

Beauty and Design

I am fascinated by the profound impact nature's dynamic movement has had on my wellbeing – on my peaceful state carrying through my most profound grief. I have focused on the actual properties of the distraction and questioned in my day-to-day job how my team might recreate them in our interiors. Building design is complex, and where and when to place these dynamic elements is not arbitrary. Interior designers must not only get building occupants to pause and orient themselves toward distracting details but also move clearly through complicated floor plans that often can be confusing and chaotic. A portion of our professional skill is tied to this very effort of simplifying wayfinding. Have you ever tried to find your way through a hospital while you were in a disturbed state? It is hardly peaceful. If I may borrow an image from my sailor husband, it is like being lost at sea without a rudder. You survey the environment, and if the dynamic energy of the environment is not sending you clues, then your experience is flat, like a sailboat without wind. Yet we, as humans, are required to navigate many facilities with neither rudder nor power. How can interior design power a journey when there is no dynamic energy?

Planning is a design process that creates a visitor's intuitive rudder. Planners are key, as they too can design into a building floor plans personal wellbeing and harmony at a variety of levels. Cal Newport's book *Deep Work* (2016) illustrates this design concept as he describes the Eudaimonia Machine, a machine that takes its name from an ancient Greek concept of eudaimonia, a state in which one achieves full human potential. In the book, a conversation with an architectural professor, David Dewane, describes the machine as a building with five layers and without hallways, so that "one does not bypass any of the spaces as one goes deeper into the machine" (p. 95). The building creates a setting where users "get into a state of deep human flourishing" (p. 95). My firm, CAMA, has used this concept within a biophilic context. It is used to help our clients understand how their building systems can move away from complex corridor-centric mazes. The illustration in Figure 5.6 is used in all our planning meetings to prepare decision-makers to break from conventional wisdom about how to transform interior journeys. Like Newport's architectural dialogue, the deeper one goes into the Eudaimonia Machine, the more communal it is, and the more one flourishes. Unlike Dewane's description, CAMA labels the spaces not by function but rather by outcome, as these are the outcomes Maslow sought to deliver fulfillment. Maslow's hierarchy (Figure 5.7) defines the outcomes, moving from human survival to

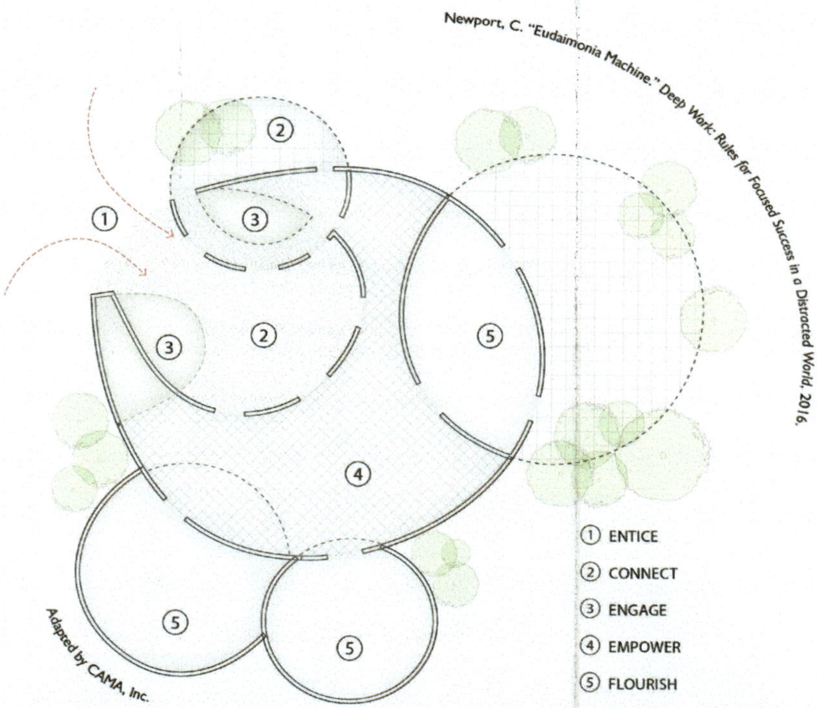

FIGURE 5.6 CAMA's experience map of the Eudaimonia Machine.

Source: Illustration by CAMA.

flourishment (Maslow & Lewis, 1987). As the ultimate fulfillment, flourishment can also be defined as the enjoyment of life achieved in the eudaimonia experience in Sartwell's triad of truth, goodness, and beauty, and in Gawande's wish to be alive.

The Eudainomia Case Study

One does not often have the opportunity to align an institutional vision with planning design concepts or design interventions, such as movement and biophilic design. Recently, CAMA, in collaboration with the Pomarico Design Studio, designed an innovation center for the Gaylord rehabilitation hospital. This project serves as an example from which to gain insights that might empower others to focus similarly on peace for a life indoors. The design was based on the application of the dynamic design principles of movement I experienced at my Peace House, and much of the project planning was done using the layered principles noted in the Eudaimonia Machine: enticement, connection, enlightenment, empowerment, and flourishment. Aligned with

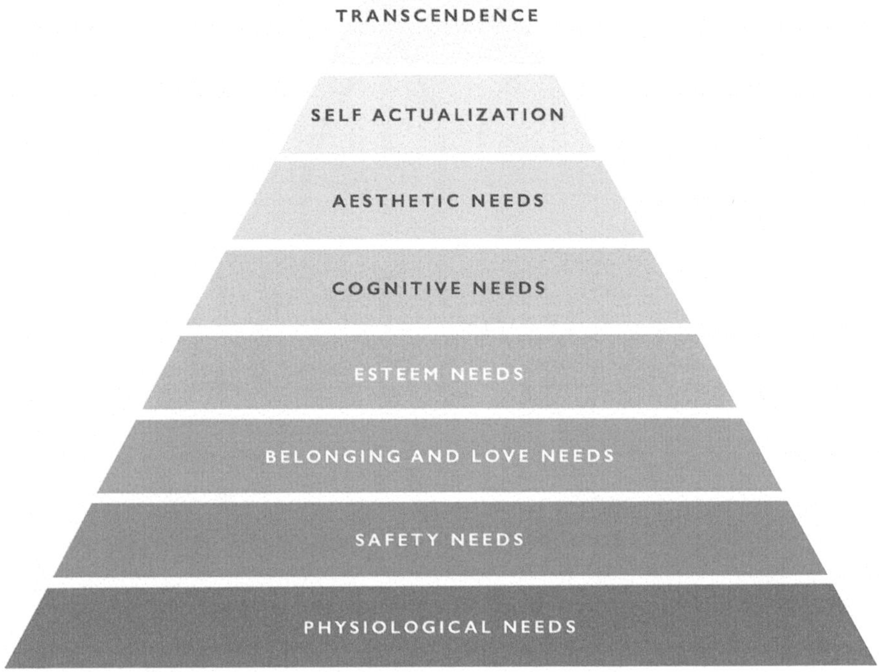

FIGURE 5.7 Maslow's hierarchy of human needs.

Source: Maslow and Lewis (1987), graphic by CAMA.

the hospital facility's masterplan, the rehabilitation community was to focus on a series of enabling projects, each designed to enable those with physical and cognitive challenges to accept, manage, and improve functional loss by utilizing nature and engaging their mobility and sensory functions.

The first enabling project was to house a new education and research team, including an integrative medicine practice. These new programs were designed to incubate practice growth with the hope of improving patient lives so that compromised patients would not just survive injury but also enjoy life. Nursing education and the residency program were to share contemplative lounges and simulation educational classrooms. The innovation center was to share a floor in the outpatient services building, enabling patients and their families to engage and interact with researchers and the clinical staff. The main entry to this building was to serve as the main entry to the entire facility until the next phase of enabling projects was completed.

Enticement

Healthcare designers have reason to always use the principle of enticement, as there is an incentive to build patient satisfaction in care delivery. Satisfaction is measured through required patient surveys, and satisfactory scores become significant since reimbursement in healthcare is often compensated based upon national rankings. Generally, an expectation about an environmental experience is said to become reality in the first six feet up the drive and the first sixty seconds in the door. If the design of those areas provides a transformative experience for visitors to the site rather than just a transactional one, then it is likely that the user's perceptions will exceed their expectations. It is there where the memory of the experience becomes important for user surveys (Becker & Douglass, 2008). When healthcare services are complex, the user perceives only those aspects which are familiar or unique. So, transformative experiences in the first sixty seconds in the door are more frequently remembered and reported and thus become enormously important (Berry, 1995).

Enticement is all about attraction, like a sunrise or a sunset that offers a colorful refraction of light. Enticement at the Gaylord project begins at the building's entry, as it takes those with limited mobility across a catwalk mid-section of a three-story opening connecting at its main level. The building has two three-story light wells, and the lower level of the Gaylord renovation project took advantage of them by articulating them with multi-story chandeliers that simulate the glint of sunlight by day or the twinkle of stars at night. The sparkle that comes from drops of lighting entices those who pass by but also those who enter and those who are on other floors, causing them to pause, and captivating many who are particularly bound by physical and cognitive challenges. It is here where the inspiration from observing nature's movement influences product choice. The light is not merely illumination but rather a positive distraction. Its sparkle catches the visitor's attention, entices them to pause and forget, for a moment, their injury or disease. During construction, we have seen patients pause to be mesmerized by the sparkle and glint of the light from a fixture that is suspended in the unarticulated void flanking the three-story wells. Delight in positive distractions offers moments of respite for personal wellbeing (Ulrich, 1992). This enticement helps reduce the fear of the unknown for those who enter and are about to confront the challenges dealt to them or a loved one. This effect is similar to the dynamic movement of the sea in masking my grief. Philosophically, Elaine Scarry (1999) defines the opposite of beauty not as ugliness but rather as injury. In her book *On Beauty and Being Just*, Scarry notes, "Matisse never hoped to save lives. But he repeatedly said that he wanted to make paintings so serenely beautiful that when one came upon them, suddenly all problems would subside" (1999, p. 33). When the visual clues that exceed expectations contain beauty, then the hope of that which has been taken away might be restored. Therefore, at the entry of an interior experience, one needs to be peacefully enticed to enter in a personal state of wellbeing and connect with others in a building in a state of harmony. A visitor must be cued to restore beauty, the opposite of injury. Their experience, as they enter a space, must prime them to receive peace. The way one is received is not accidental but rather about a gracious ceremonial moment of acceptance.

Connection

The next level of eudaimonia, connection, indicates that something must allow people to form a link with their surroundings. In cities and complex buildings, we call these landmarks. The best city planners and wayfinding experts use them in noting geographical locations, pride of place, or legacy in community. Landmarks must be relevant to make them memorable. At the entrance to the Women's Cancer Clinic at the Smilow Cancer Hospital, where the owner originally wanted to put a directional sign, a show-stopping painting of a bouquet of flowers, "Dahlias," created by artist Cynthia Packard, was placed to comfort women facing an uncertain future. This was subliminal messaging that created personal connections and clear wayfinding. A visitor seeks prospect (an understanding of direction) and refuge, such as a protected place from which one watches a positive distraction away from the pain and discomfort often associated with healthcare (Heerwagen & Gregory, 2008).

What connects us to a particular space? Does a biological need to survive force a connection to a tribal space? Or is it the color or a bloom that exists in a spot in nature, as when monarch butterflies flock to blooming goldenrod? What is it about the design of an environment that attracts us and provides communal interaction, opening pathways to connection? In these gathering spaces, like parks in our cities, using simultaneous perception, Hiss suggests that we might be able to "discover what we have in common with all of the life around us" (1990, p. 48.)

Once I was asked by a mother of a hospitalized child to give her "direction and diversion – Get me to where I have to go and make me forget why I am going there." These are the connections interior designers must make to set the stage for a peaceful journey as a building interior unfolds the occupant's journey. "In nature there is much to discover by looking closer" (Heerwagen & Gregory, 2008, p. 231), so give all visitors time to pause and look closer; it is here where the dynamic motion mesmerizes.

Enlightenment

Enlightenment is the third layer of eudaimonia. The daily and seasonal cycle of light is also needed indoors as full-spectrum light, not the typical lamping of one temperature that does not interact with human biorhythms. The aligning

FIGURE 5.8 Entice: A patient's entry experience must deliver communal harmony. Lobby entry of Gaylord Specialty Healthcare in Wallingford, CT (Interior Design: CAMA, Inc., Architect: PDS, Lighting Fixture: Lodes).

Source: Caryn B. Davis Photography.

of daytime circadian rhythms can be achieved with electrical lighting control systems. A broader discussion with engineers, lighting designers, and light control manufacturers is needed to push for more modulation in interior lighting levels that are efficiently and effectively aligned with our body rhythms. Light therapy is particularly important for aging patients and patients whose sleep patterns are interrupted by treatment or pain.

Natural light is moving all day long, in different intensities seasonally, and it dances with the movement of clouds and openings in the canopy of trees and bounces off reflective objects. Interior designers can borrow from nature by casting shadows and dappled effects that are found in non-conventional sources to achieve the effect of a stroll outdoors, using natural daylight from floors above, using indirect lighting to illuminate reflective surfaces, and

FIGURE 5.9 Connect: Giving visitors time to pause and look closer for personal wellbeing. "Dahlias" oil painting by Cynthia Packard, Smilow Cancer Hospital at Yale-New Haven, New Haven, CT, Healing Arts Program by CAMA.

Source: Rick Scanlan Photography.

modulating light levels. Modulated light levels can be offset with brightly lit landmarks, and using landmarks can be a lovely way to draw a person into a building once enticed to enter. It can connect them to a relevant moment that may trigger a memory and then prepare them for the next layer of their journey. Scrutiny of all engineered sensory perceptions is needed, so that they can be understood and used in standard design applications.

Enlightenment arrives not only through vision but also by sound. Bluetooth technology can add sound easily and customize it with artificial intelligence that will prompt sounds when a smart signal passes by. Electronic controls coupled with artificial intelligence offer enticing events when landmarks are not enough and customized messaging is needed. There is little being done with sound in the movement of people through environments. I delight in the sound one hears as one shuttles between United terminals at O'Hare Airport in Chicago; it makes a long arduous walk delightful and energizing. Although music is used in therapies, it is rarely used in wayfinding. Healthcare environments are full of complex sound. Hildebrand (2008, p. 265) says that we are genetically programmed to respond positively to complexly ordered sound (music) but not to chaotically complex sound (noise). Connecting with one's environment can be fun and whimsical; it can also be serious and thoughtfully considerate and enjoyable. Imagine the possibilities. Design to entice (nudge) and enlighten your audience to connect with their own journey, not another's. Make it engaging and enjoyable.

The eudaimonia level of enlightenment requires spaces for people to gather and interact – places where more attentive behaviors can occur simultaneously in communal harmony. In our case study, the rehabilitation hospital's newly designed lower level is surprising to those who arrive, as it is made up of directional sinuous pathways that allow all to interact. It has a mix of patient, clinical, and support staff areas. In the planning of the environment, the design team committed to following the master plan values to transform care at this specialty care hospital. The design interventions utilized materials and building systems that allowed the interior to replicate the experience found outdoors in the 400-acre hilltop site this facility sits upon. The interior environment replicates the feeling found navigating the campus grounds by incorporating curved walls as intersecting planes made of a variety of surfaces. Like the outdoors, important cues are found in vertical wooden wall lattice connected by reflective glass surfaces with variegated color in natural patterns, creating contemplative landmarks that are memorable.

Empowerment

Once enticed, connected, and enlightened, one is ready to be empowered to engage in communal harmony. In order to empower people, the environment must engage them. Think about a walk or drive through nature shared with a companion. Sacred spaces that capture attention and focus one's entire being on the majestic confluence of total sensory immersion, and long calming views of nature provide little or no threat. In environments like these, people come together in harmony and are likely to engage with each other, promoting empowerment. At the Gaylord project, pathways, not corridors, were designed and dotted with perches to stop in one's journey. They were created to encourage pacing while pondering; they lead to a central gathering space in the form of a round kiva. Hoteling spaces for visitors surround the kiva. These spaces, where serendipitous gatherings can occur, can be intriguing and become a conversation starter for open collaboration and empowerment.

Flourishing

Once empowered, then and only then can one flourish with one's whole physical, neurological, psychological, and spiritual self present. Flourishment ignites enjoyment. Enjoyment in the built environment leads to peacefulness and harmony. As we thrive and flourish in nature, so can we indoors. CAMA's hope

FIGURE 5.10 Enlighten: Directional sinuous pathways that allow all to interact. The Institute for Advanced Rehabilitation, Gaylord Specialty Healthcare in Wallingford, CT (Interior Design: CAMA, Inc., Architect: PDS).

Source: Caryn B. Davis Photography.

is that Gaylord's Innovation Center will stand as an example of a new way of using what is already in our designer toolbox so that we can redefine and measure transformational interior experiences. It can help lay the foundation for providing prospect and refuge indoors in settings that encourage us to thrive and flourish. Many research projects should follow so that we can learn from this project's ability to deliver peacefulness both in user wellbeing and in communal harmony (see Figure 5.12).

The Path Forward

Throughout these chapters there exists a baseline of knowledge that gives permission to those who will lead and bring their best professional and leadership skills to activate a building occupant's behavioral change. These permissions build the opportunity to imagine a solution for this primal unmet need: peace or individual wellbeing in communal harmony for populations that live predominantly indoors. In the blurred demarcation line of outdoors and indoors lies a human being that has evolved to respond to external triggers. Further study is needed to understand how to promote those triggers towards peaceful experiences indoors. The beauty of this quest is that interdisciplinary design disciplines have an opportunity to participate and co-design.

FIGURE 5.11 Empower: A sacred space with a majestic confluence of total sensory immersion. Hope Chapel, Gaylord Specialty Healthcare in Wallingford, CT (Interior Design: CAMA, Inc., Architect: PDS).

Source: Caryn B. Davis Photography.

FIGURE 5.12 Flourish: Kiva to pause or gather. Milne Institute for Research and Innovation, Gaylord Specialty Healthcare in Wallingford, CT (Interior Design: CAMA, Inc., Architect: PDS).

Source: Caryn B. Davis Photography.

This effort can fuel design advocacy organizations through alignment with an organization like the WHO and find purpose in delivering an expanded base of multi-disciplinary knowledge armed to deliver peace by design.

A definition of Peace for Life Indoors might be "A dynamic interior environment, designed and detailed so that one is able to be enticed, connected, enlightened, empowered and flourish through physical, neurological, psychological and spiritual evidence-based triggers resulting in personal wellbeing, harmony or peace." An opportunity for peace for life indoors is soon to come.

Peace!

References

Becker, F., & Douglass, S. (2008). The ecology of the patient visit: Physical attractiveness, waiting times, and perceived quality of care. *The Journal of Ambulatory Care Management*, *31*(2), 128–141.

Berry, L. (1995), Relationship marketing of services—growing interest, emerging perspectives. *Journal of the Academy of Marketing Science, 23*, 236–245.

Bolinger, R. J. (2020). Varieties of moral encroachment. *Philosophical Perspectives, 34*(1), 5–26.

Gawande, A. (2014). *Being mortal: Medicine, and what matters in the end*. Picador.

Gladwell, M. (2000). *The tipping point*. Little, Brown & Company.

Gooley, T. (2017). *How to read nature: Awaken your senses to the outdoors you've never noticed*. Experiment Publishing.

Heerwagen, J., & Gregory, B. (2008). Biophilia and sensory aesthetics. In S. R. Kellert, J. Heerwagen, & M. Mador (Eds.), *Biophilic design: The theory, science, and practice of bringing buildings to life* (pp. 227–241). John Wiley & Sons.

Hildebrand, G. (2008). Biophilic architectural space. In S. R. Kellert, J. Heerwagen, & M. Mador (Eds.), *Biophilic design: The theory, science, and practice of bringing buildings to life* (pp. 263–275). John Wiley & Sons.

Hiss, T. (1990). *The experience of place: A new way of looking at and dealing with our radically changing cities and countryside*. First Vintage Books.

Institute of Medicine. (2001). *Crossing the quality chasm, A new health system for the 21st century*. National Academy Press.

Maslow, A., & Lewis, J. K. (1987). Maslow's hierarchy of needs. *Salenger Incorporated*, *14*(17), 987–990.

Merriam-Webster. (2023). Peace. In F. C. Mish (Ed.), *Merriam-Webster*. https://www.merriam-webster.com/dictionary/peace#:~:text=Legal%20Definition-,peace,for%20by%20law%20or%20custom

Naylor, R., & Sunstein, C. (2008). *Nudge, Improving decisions about health, wealth, and happiness*. Yale University Press.

Newport, C. (2016). *Deep work: Rules for focused success in a distracted world*. Grand Central Publishing.

Prum, R. O. (2018). *The evolution of beauty: How Darwin's forgotten theory of mate choice shapes the animal world – and us*. Doubleday.

Sartwell, C. (2004). *The six names of beauty*. Routledge.

Scarry, E. (1999). *On beauty and being just*. Princeton University Press.

Taylor, J., & Wacker, W. (1997). *The 500 year delta: What happens after what comes next*. Harper Collins.

The Center for Health Design (n.d.). *The Pebble project*. https://www.healthdesign.org/sites/default/files/Pebble%20Project%20Brochure.pdf

Ulrich, R. S. (1992). How design impacts wellness. *Healthcare Forum Journal, 20*, 20–25.

US Department of Health & Human Services. (2023, May 3). https://www.hhs.gov/sites/default/files/surgeon-general-social-connection-advisory.pdf

Wacker, W., Taylor, J., & Means, H. (2000). *The visionary's handbook.* Harper Business.

Wilson, E. O., & Kellert, S. R. (1993). *The biophilia hypothesis.* Island Press.

World Health Organization. (2023). https://www.who.int/data/gho/data/major-themes/health-and-well-being#:~:text=The%20WHO%20constitution%20states%3A%20%22Health,of%20mental%20disorders%20or%20disabilities.

World Health Organization. (n.d.). https://www.who.int/initiatives/who-health-and-peace-initiative#:~:text=Health%20and%20peace%20are%20interrelated,cannot%20be%20peace%20without%20health%E2%80%9D.

6

EVERYDAY PRODUCTS AS A MEANS OF FOSTERING PEACE, MEANING, AND HAPPINESS

JungKyoon (Jay) Yoon

Take a moment to answer the following two questions: (1) what are the things that bring you pleasure but fail to make you happy? (2) What are the things that don't necessarily bring you pleasure but still make you happy? I assume your answers to each question may refer to different things.

For many decades, pleasing consumers has been of central concern to designers, including myself, focusing on capturing consumers' needs and wants. Accordingly, design research has generated a wealth of knowledge about how to design products that stimulate pleasurable experiences. The term "product" refers to anything that needs to take on a form. While, in practice, it is commonly associated with physical objects like consumer products, the term can also encompass designed spaces, software, product-service-systems, or environments where these elements function as cohesive units (Erlhoff & Marshall, 2008). Knowing how to make something attractive, efficient, and easy to use has helped designers make products that make consumers feel good, thereby selling well. However, a pleasurable experience does not always necessarily foster peace or serve a worthwhile purpose. Traditionally, design has not paid enough attention to differentiating the answers to the two aforementioned questions in its effort to make a positive impact.

While it may sound a bit naïve, I began my career as an industrial designer with a clear purpose: *making a positive impact through design*. Looking back at my previous work, regardless of the product types, encompassing consumer electronics, interactive installations, social media, Internet search interfaces, and so on, I must admit that I heavily focused on creating things that almost exclusively fall under the first question. In hindsight, my design skills and knowledge, honed over years of training and practice, were geared towards addressing how to make product experiences more pleasurable as a means of convincing

DOI: 10.4324/9781003462019-16

consumers to buy more and want more, increasing the number of consumers, enticing consumers to spend more time with the products, and many similar intents. Overall, my work was meant to promote product uptake, and in this effort, while design successfully served as a source of competitive advantage that drove sales, it did not take a long time for me to recognize a mismatch between my initial motivation and the real impact of the things I designed. A nagging question kept lingering in the back of my mind – Did I want my family and friends to live in a world filled with the things I had designed so far? The answer was a resounding no, and for the past decade, this realization has driven me to seek different ways of designing than I was used to or simply good at.

A new possession, such as a smartphone, water kettle, or a speaker, can bring moments of intense happiness. This mode of happiness, however, is fleeting rather than sustainable because people adapt to the pleasurable experiences. Even large personal purchases, like a new laptop computer, can lose their initial appeal over time. While a new high-performing computer may bring happiness initially, the positive experiences (such as touching its smooth surface while unpacking, seeing an aesthetic interface on its high-resolution display) and emotional responses (such as fascination and pride) decrease as the user adapts to owning it (Sheldon & Lyubomirsky, 2012). This phenomenon is not limited to specific products but is a general tendency in humans to adapt to new situations and stimuli, leading to diminishment in happiness.

> Most human beings have an almost infinite capacity for taking things for granted. By the mere fact of having come into existence, the most amazing novelty becomes in a few months, even a few days, a familiar and, as it were, self-evident part of the environment.
>
> *(Huxley, 1950, p. 69)*

As illustrated previously, over time, we adjust to changes in our circumstances and no longer experience the same level of delight, excitement, calmness, or peacefulness from them, a phenomenon referred to as hedonic adaptation (Lyubomirsky, 2011). Hedonic adaptation also occurs in our daily product consumption and usage, including brand-new home appliances, exciting social media platforms, and abundant music available on streaming services. When design is solely focused on sparking pleasure, people quickly become accustomed to the pleasure elicited by a product, eventually finding it mundane. Eventually, their appreciation of the product fade, and they take positive experiences for granted, with the happiness effect wearing off before long. As Aldous Huxley pointed out, it is our nature to quickly take things for granted and pursue the next thing, forever chasing the same pleasure, like a rat running on a "hedonic treadmill" (Brickman & Campbell, 1971). In that sense, happiness achieved with products is considered momentary. Thus, well-being theorists have been generally skeptical about the contribution of products to lasting user happiness.

These observations raise questions about the traditional design aim of stimulating pleasure and reducing displeasure to increase users' happiness that have been dominant in the perspectives of ergonomics and usability (Hassenzahl et al., 2021). In response, a growing number of research initiatives has recently made important headway in exploring how design can willfully contribute to the happiness of individuals and communities. These initiatives include empathic design (Koskinen, 2003), positive design (Desmet & Pohlmeyer, 2013), positive computing (Calvo & Peters, 2014), and experience design (Hassenzahl, 2010), to name a few.

Although these initiatives differ in their approaches and theoretical underpinnings, they all aim to use design as an enabler, focusing on the quality of life or happiness of individuals and society by investigating life aspirations and fundamental human needs. The key argument of these initiatives is that while products may not be a direct source of profound happiness, they can still be designed to contribute to happiness by influencing positive activities and increasing awareness of them. Products can play important roles in activities that positively affect our happiness, both directly as a resource and indirectly as a source of inspiration. In the words of Desmet, "products do not make us happy, but can help us in becoming happy" (2011, p. 3).

This chapter focuses on how design can deliberately inspire and enable positive activities to promote user happiness. The key question, then, is what kind of positive activities one should engage in and how can designers contribute. To answer this question, I propose four design opportunities that enable positive activities supporting both personal and collective values. These design opportunities are centered around positive intra- and interpersonal activities, which are known to be resistant to hedonic adaptation (Lyubomirsky, 2011). The chapter begins by describing the theoretical account of "happiness" and "well-being" as an experiential impact to design for, revealing that design-mediated happiness is not solely dependent on the product itself but largely on users' activities and usage contexts. Then, the chapter introduces four activity-oriented design opportunities through the lens of social relationships and explains each one through exemplary design cases. Finally, I discuss challenges and future research directions from a methodological perspective. The hope is that this chapter will inspire design students and practitioners to approach their creative design processes with a focus on human happiness, as well as stimulating wider debates about the impacts of design on individual and collective well-being.

Design for Peace and Happiness

The term "happiness" has different meanings, both in everyday dialogue and in scientific research. Although there is no universally accepted single definition of happiness or well-being, it is generally agreed that it refers to an individual's overall satisfaction with life and fulfillment of positive functioning (Veenhoven, 2011). Here, the focus is on individuals' subjective evaluation of their lives as a whole (i.e., subjective well-being) rather than on the degree to which external conditions for having a high quality of life are met (i.e., objective well-being).

While pathways to happiness have been operationalized in diverse ways, they are grouped into two broad categories: (1) hedonic and (2) eudaimonic pursuits (Deci & Ryan, 2008). Both pursuits contribute to happiness in different ways. From a hedonic perspective, happiness is the sum of one's pleasurable moments, including peacefulness and excitement, with a focus on "here and now," maximizing pleasure and minimizing displeasure such as physical pain and psychological distress. Examples are enjoying the high-fidelity room-filling sound of a speaker and relaxing by watching an amusing film after a long day of work. From an *eudaimonic* perspective, happiness is achieved by fulfilling moral values and virtues, with a focus on long-term implications for the development of oneself or society. Examples are taking good care of the environment by consuming technology produced with less environmental harm and developing personal talents through online tutoring. These pursuits tend to have a more lasting and meaningful effect on life appreciation beyond momentary gratification (Ryan & Deci, 2001).

When it comes to the concept of "peace," its meaning can vary from the two different perspectives of well-being as described previously. From hedonic perspective, peace is understood as an emotional state of calm, tranquility, or enjoying easy and relaxing moments (Desmet & Fokkinga, 2020). Several studies have examined the relationship between peace and individual well-being, showing that higher levels of peacefulness, characterized by higher levels of calmness and lower levels of stress, were closely associated with higher levels of well-being (e.g., Mogilner et al., 2011). On the other hand, from the eudaimonic perspective, the concept of peace extends beyond individual experiences and encompasses harmonious relationships, social cohesion, and collective welfare (i.e., interpersonal peace; Diener & Tov, 2007). It is also characterized by the presence of values and virtues such as justice, equality, and compassion and the absence of violence, injustice, and social disparities within a community or society. A study by Diener and Tov (2007) on the relationship between peace and social well-being showed that countries with higher levels of peace, as indicated by higher levels of social support, nonviolence, and equality, tended to have higher levels of social well-being.

This chapter specifically pays attention to the contribution of design to eudaimonia, distinguishing the short-lived moments that are often considered moments of happiness in everyday conversation. According to Veenhoven (2011), these enjoyable moments are "pleasures" and may contribute to, but are not, happiness. Here, I use "happiness" and "well-being" interchangeably to refer to a sense that one's life is good, meaningful, and worthwhile, that is, enduring life appreciation, based on the definitions of Lyubomirsky (2008) and Dolan (2014). In the same vein, the chapter adopts the eudaimonic perspective on interpersonal peace.

Happiness can be achieved even when one's circumstances might not be objectively favorable. For example, studies have shown that people who live in poverty are still capable of pursuing higher-order needs such as peace, love, esteem, and excellence, even though they are struggling to achieve lower-order needs such as food, shelter, and physiological health (Wahba & Bridwell, 1976; Wicker et al., 1993). Therefore, designers should refrain from assuming that such higher-order needs remain unachievable until lower-order needs have been satisfied, as exemplified by Maslow's hierarchy of needs (1943). Otherwise, especially in an aging society, well-being would then only be reserved for a few (Huber et al., 2011). In line with these findings, positive psychology, which studies the science of happiness, emphasizes achieving personal and social potential, self-transcendence, and personal growth despite physical or mental health challenges (Seligman, 2008).

Two decades of research in positive psychology have generated mounting evidence supporting the idea that happiness is worth pursuing. For example, the literature reviews of Lyubomirsky et al. (2005) and Eid and Larsen (2008) show that happier people are more sociable and energetic, more charitable, cooperative, and open-minded, and therefore likely to foster peace and less likely to engage in violence. Accordingly, well-being has become a salient issue, often on many countries' political agendas as a priority to foster prosperity while safeguarding the protection of the planet. The elevation of the topic is well illustrated by the fact that the creation of the United Nations (UN) 2030 Sustainable Development Goals (SDGs) involves 17 goals for health and prosperity (United Nations [UN], 2016).

In addition to research efforts in positive psychology, design as a discipline can also play a crucial role in realizing the vision of a good life. It can enable and encourage activities that lead to well-being and thriving communities. The combination of emerging technologies with physical environments, systems, services, and artifacts provides numerous possibilities to make a positive contribution to people's well-being. These possibilities can be integrated into their daily lives beyond traditional clinical therapeutic settings. Acknowledging these potentials, design for well-being is now considered one of the grand worthwhile challenges in design and human–computer interaction (Stephanidis et al., 2019). This chapter embraces the perspective of Desmet and Pohlmeyer (2013) on design for well-being, which aims to help people function well psychologically and live well. This involves more than just occasional feelings of peacefulness and happiness and, crucially, sees individuals and communities as creators and managers of their own good life, rather than simply as consumers.

Positive Activity as the Objective of Design

One of the core findings of well-being research is that happiness is determined much less by what we own than by what we do, which resulted in the well-known "activity advice": to achieve sustainable increases in happiness, it is more effective to cultivate intentional positive activities than to change one's circumstances or acquire new things (Lyubomirsky et al., 2005). The activity advice fits with the experience recommendation voiced in consumer research, which states that if you want to become happier, buy life experiences instead of material items. Numerous studies have shown that experiential purchases provide more long-lasting happiness than material purchases. Essentially, both recommendations advocate that spending money on positive activities like hiking, trying a new restaurant, or taking a dance class is a better investment in well-being than spending money on material possessions that are susceptible to hedonic adaptation (Sheldon & Lyubomirsky, 2006).

Plausible explanations of these observations are that experiences are more open to positive reinterpretations, form a more meaningful part of one's identity, and contribute more to social relationships (Van Boven & Gilovich, 2003). Pursuing "activity" goals, such as cooking a new recipe or participating in a movement, allows a person continually to experience new challenges, take on new opportunities, and fulfil diverse psychological needs (Layous & Lyubomirsky, 2014). These activities can be both mental, like thinking about pleasant memories, and behavioral, like going for a bike ride.

These empirical and theoretical findings suggest that well-being driven design should focus on products as enablers of positive activities. The design process needs to focus on identifying effective and relevant positive activities in the intended contexts and determining how such activities can be integrated into human-product interactions. In fact, whether intended or not, many of our daily activities are inherently shaped by products: activity is inseparable from materiality (Dourish, 2001), and humans act with technology, with each activity inevitably shaped by the opportunities and limitations imposed by the designed products (Verbeek, 2011). For example, our exercise routines depend on the equipment we use, and our productivity relies on the software installed on our computers (Yoon & Kim, 2022).

Imagine using a smart watch to track your fitness progress during a hike with your friends. When we think back to this experience years later, it is unlikely that we first recall the watch's features. Rather, we would recall the beautiful scenery, the feeling of accomplishment, or the memories with our friends. While the watch is an essential part of this experience, it may not be the key source of happiness. If designers only investigate the direct effects of handling a product, they may miss out on the most profound opportunity. By acting as a resource that inspires or enables meaningful activities, design can

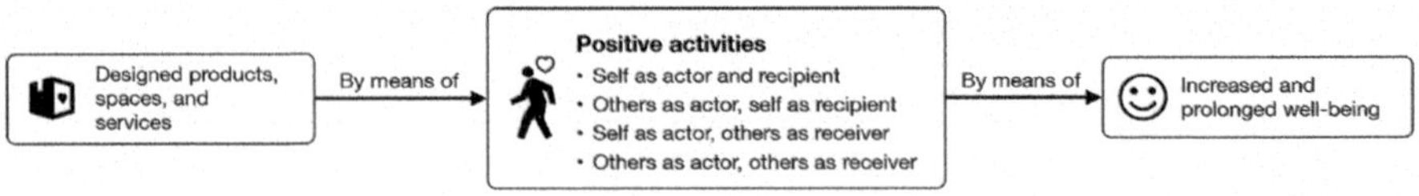

FIGURE 6.1 A scheme of design-mediated positive activity that affects user happiness.

indirectly contribute to our well-being. Additionally, as symbolic representations, design can also bring us towards positive aspects of our lives or remind us of meaningful experiences (Casais et al., 2018). These experiences should become explicit objectives of design, such as a smart bike that encourages users to explore new routes (see Figure 6.1).

There are two ways activities can be facilitated by products: First, existing products whose main function is not to support positive activities can be improved with particular features, for example, a social media platform that encourages users to share respectful messages. Second, new products can be developed specifically to foster a particular activity that increases users' happiness, for example, an interactive toy that trains children to be more mindful. In *Positive Computing* (Calvo & Peters, 2014), the former is called "Active Design" and the latter "Dedicated Design." In both cases, the product itself is not the direct contributor to increasing user happiness but rather promotes it indirectly by supporting certain positive activities. Therefore, this chapter builds on activity-oriented approaches that can be integrated into human–product interactions.

Four Activity-Oriented Design Opportunities

Designing for well-being by facilitating positive activities through products is a multilayered endeavor. It involves choosing an activity, a context for the activity to take place, and a product, which plays a key role in the activity and acts as a mediator (Verbeek, 2011). To this end, this section introduces four activity-oriented design opportunities based on the model of Revord et al. (2017) developed based on the analysis of positive psychology interventions in social contexts.

A person's subjective well-being is not only determined by the extent to which their individual needs are fulfilled but also the degree to which the needs of others and society are met (Tay & Diener, 2011). Social relationships contribute to well-being through social integration, which involves engagement in various social activities or relationships (Cohen, 2004). For example, social status, which is the respect and admiration people receive from their groups (e.g., neighbors or co-workers), is an important predictor of well-being (Anderson et al., 2012). Similarly, the happiest individuals all reported strongly positive interpersonal relationships (Diener & Seligman, 2002). Given the importance of social interactions, I view design-mediated happiness-enhancing activities through a social psychological lens, meaning that I consider the impact of the real, imagined, or implied presence of others on the efficacy of design. In my view, this viewpoint could serve as an entry point to promote peace and the well-being of individuals and communities.

Every social interaction involves different stakeholders, each having unique experiences of the situation. For instance, an act of kindness can potentially affect three parties: the actor, the recipient, or a non-involved witness. Therefore, when designing positive activities, it is essential to start with the following questions: Who is the actor performing the action? Who is the recipient getting the action? And whose outcome are we focusing on? (see Table 6.1). In this perspective, there are four broad objectives of design: (1) intentionally engaging in positive activities (self-as-recipient), (2) being induced by others to engage in positive activities (other-as-actor, self-as-recipient), (3) influencing others to engage in positive activities (self-as-actor, other-as-recipient), and (4) being influenced by others who are engaged in positive activities with someone else's well-being in mind (other-as-actor, other-as-recipient).

Differentiating the roles of users as actors and recipients of positive activities can support designers to influence users' well-being in a systematic way. In a design process, perhaps, it can help designers broaden their view regarding

TABLE 6.1 Examples of design in each of the four design opportunities

		Actor	
		Self	*Other*
Recipient	Self	**1. Self–Self** • Using a guided meditation app to reduce stress before a big presentation • Using a fitness tracker to monitor and boost positive emotions	**2. Other–Self** • A loved one sending a motivational text message to support a healthy habit change • A therapist recommending a mental health app to a patient to help with anxiety or depression
	Other	**3. Self–Other** • Sharing an uplifting post on social media to support friends or followers going through a difficult time • Organizing a virtual event to raise awareness of emotional well-being	**4. Other–Other** • Admiring a social activist who uses social media to raise awareness for a cause • Being inspired by a volunteer who organizes an event to support local small businesses

Source: Adapted from Revord et al., 2017.

what kinds of activities would be appropriate for users to engage in and identify how design can support the activities. Further, being more specific in the ways of supporting users as actor and recipient may stimulate designers to inquire about users' and other stakeholders' needs, capabilities, and concerns in detail. The following section describes each of the four design opportunities with design examples and relevant design considerations.

Opportunity 1: Self as Actor and Recipient

The first opportunity is at the intersection where the self is both actor and recipient. It involves positive activities in which individuals act upon themselves to enhance their own happiness. Examples of positive activities that fall within this quadrant include intrapersonal prosocial behaviors like self-spending and self-compassion. For instance, one can listen to calming music to reduce personal anxiety, practice mindfulness to increase awareness of one's emotions, or engage in physical exercise to uplift positive emotions.

Designers can explore how products can be designed to support particular positive activities with the intention to stimulate personal growth, intimacy, self-efficacy, and physical health. Most self-administered design such as behavioral intervention technologies (BITs) involve the self, acting on the self, such as savoring, positive reminiscence, and acting kindly towards oneself (Slovak et al., 2023). This refers to products that have been redesigned to support self-directed activities to the conventional design (i.e., active design), as well as products that have been created with the main purpose to be resources of positive activities in this opportunity (i.e., dedicated design).

An example is Purpal (Yoon et al., 2021), an interactive device designed to improve how people regulate their emotions when making purchases (see Figure 6.2 left). It addresses issues like buyers' regret and hedonic adaptation

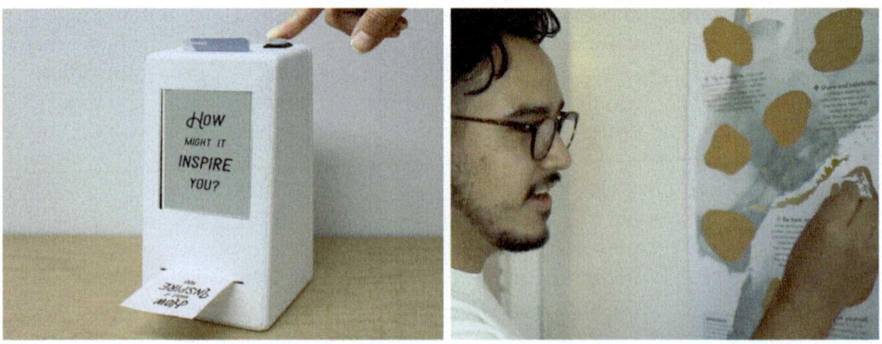

FIGURE 6.2 Design cases that fall under the first opportunity: Self as actor and receiver. Purpal designed by Shuran Li (left image), Revealing Moments designed by Jeremy Faulk (right image).

by asking users adaptive questions about their purchase intentions. These questions serve three main purposes: (1) helping users think about the experiences the purchase will provide, (2) encouraging them to share positive experiences with others, and (3) linking the purchase to their personal goals or aspirations. For example, if the purchase is a gift, a question could be "What message do you want to convey when you give it?" If it is sports equipment, the question could be "How would this item contribute to your mind or body?"

Revealing Moments (Faulk & Yoon, 2023) is an example of a product that allows individuals to cultivate their own happiness (see Figure 6.2 right). It is an interactive poster designed as a scratch-off poster, aiming to encourage users to appreciate their positive experiences using four savoring techniques: absorbing, anticipating, self-congratulating, and sharing with others. Unlike the common approach of prescribing positive behaviors (such as keeping a gratitude journal), this poster provides a flexible starting point for savoring practices. It also offers opportunities for users to combine their chosen actions with specific savoring techniques. The scratch-off areas hide open-ended prompts (e.g., "Secret dance" or "Bring light"), inspiring users to visualize and engage in self-selected positive behaviors in their daily lives. Users have the freedom to choose activities that align with their own resources, goals, and preferences.

Notably, many of the self-oriented activities that enhance happiness in a reliable and lasting way involve treating oneself as kindly as one might treat another person. One example of such activities is practicing self-compassion. Studies have shown a positive and strong correlation between self-compassion and life satisfaction, peace, optimism, and positive affect and a negative correlation with negative affect (Neff et al., 2007). Note that the design focus needs to be on self-care, or working on personally significant goals or virtues, rather than solely on hedonic pleasure, such as temporal excitement elicited by self-directed spending. One way that spending on the self improves happiness is when people invest in their experiences. Anticipating experiential purchases, as shown in the case of Purpal, is associated with higher levels of happiness than waiting for material objects, like a luxury bag (Kumar et al., 2014). While people often assume that spending money on themselves will make them happier than spending on others, research suggests that prosocial spending is much more beneficial. This is further discussed in the third design opportunity.

Opportunity 2: Others as Actor, Self as Recipient

The second opportunity involves positive activities in which another person is the prosocial actor, and the "self" benefiting from the activities is the recipient. For instance, interventions could include receiving a compliment or having someone offer to help with a task.

Designers can explore how products can be designed to encourage other people to engage in these positive activities that benefit the intended users. This includes activities in which other people play a supporting role in helping the intended users adopt new habits or attitudes that can ultimately enhance the users' happiness. In this scenario, the focus of design is on developing a means that inspires and empowers others to share their resources (e.g., time, energy, knowledge, and materials) effectively. These "other people" could be directly related to the intended users, such as caregivers, family members, or friends, or strangers who share an interest in supporting the intended users, such as volunteers in an education program.

Be My Eyes (2023) is a mobile app designed for visually impaired individuals (i.e., receiver) to receive help in recognizing objects and navigating everyday situations (see Figure 6.3 left). The app connects the users in need with volunteers (i.e., actor) via live chat, enabling them to receive assistance with small tasks. These tasks can range from reading expiration dates on food packaging to describing the colors of clothing items. To utilize the app, a visually impaired person starts a live stream using their smartphone camera. This allows volunteers to provide assistance by describing objects, guiding camera movements, reading instructions, or helping with spills. By leveraging the ability of volunteers, the app empowers visually impaired individuals to manage their routines.

Practicing kindness and gratitude is a positive activity that has been extensively studied. Research has shown that individuals who are the recipients of kindness and gratitude can benefit emotionally and tangibly from receiving a kind act, such as much-needed help from a co-worker. Those who received small expressions of gratitude, such as "Thank you so much! I am really grateful," after helping someone were more likely to offer aid to the same person, a different person, or an organization a second time. These positive effects were

FIGURE 6.3 Design cases that fall under the second opportunity: Others as actor, self as receiver. Be My Eyes application (left image), Praise feature on Microsoft Teams (right image).

found to be mediated by increased perceptions of social worth and feeling valued by the person expressing gratitude (Grant & Gino, 2010).

For instance, Microsoft Teams, an enterprise social media platform, offers a praise feature that allows colleagues (i.e., actors) to express their recognition and appreciation for one another's contributions (i.e., receivers). This feature serves as a platform for expressing gratitude, celebrating achievements, and fostering a positive work environment in the long term (see Figure 6.3, right). Through textual and pictorial remarks, colleagues can publicly praise contributors, increasing the visibility of their work and boosting morale within the organization. Examples of praise may include statements such as, "I want to express my appreciation to the entire sales team for their outstanding performance this quarter" or "Their efforts have greatly contributed to our success."

As such, expressing gratitude requires a means to recognize other individuals as the recipients and determine to make use of their resources. Here, designers' focus is on identifying contextual, physical, cognitive, and social factors that enable or hinder the positive activities, addressing the following questions: (1) why is the activity meaningful and enjoyable? (2) what knowledge and skills do the activity require? and (3) what product and other resources are needed for the activity? For example, the activity "offering help to colleagues while working from home" that fulfills the needs of relatedness and impact would require a messenger app to initialize the activity, and the user would want to know if colleagues are free to chat before calling or texting.

The first two design opportunities promote the happiness of the self. The focus of the design is on intentionally differentiating the agent of positive activities as either the self or others and providing appropriate products as enablers of the activities. These two opportunities require designers to consider the characteristics of the self and others, such as their physical and cognitive capabilities in usage contexts. That being said, in these two opportunities, the self, as the recipient of positive activities, becomes the primary focus of attention when assessing the effectiveness of the products. The next two design opportunities focus on others as the recipients of positive activities.

Opportunity 3: Self as Actor, Others as Receiver

The third opportunity involves how doing something to benefit others (i.e., the recipient) enhances the doer's happiness (i.e., the actor). Volunteering is an example of a behavior directed towards others that has consistently been correlated with higher levels of well-being (Wheeler et al., 1998). Other examples include feeling fulfilled and rewarded after helping a stranger in need, experiencing a sense of accomplishment after mentoring students, or feeling fulfilled after donating to a charity. As such, there are multiple pathways by which committing to prosocial activities benefit the well-being of the self (i.e., the actor). One prominent pathway is that acts of kindness serve to strengthen one's place in a social network (e.g., with friends and family). In one study, children who performed acts of kindness experienced larger increases in peer acceptance than children who engaged in an alternative positive activity (Layous et al., 2012).

Chooz is an interactive gumball dispenser designed to encourage a user (i.e., actor) to engage in prosocial behaviors by stepping outside of their comfort zone. The dispenser features two chutes where balls fall and land in a basin below (see Figure 6.4 left). The flashing lights indicate that a challenge is ready for the user to interpret and act upon creatively. Each ball represents a different aspect of the challenge. One ball represents an activity, such as striking up a conversation or volunteering for a cause. The other ball represents a person or group, such as a friend, family member, or homeless shelter (i.e., receiver). The user has the option to customize the activities and people/groups associated with the balls to create challenges that are both challenging and manageable for themselves. Once a user completes a challenge, they can place the balls on Chooz as a symbol of their accomplishment.

Note that the recipient can be both human and product. People can care for another being, including products, as a way to care for themselves. For example, Lee et al. (2019) showed that just as giving care to another human can increase one's self-compassion, caring for a chatbot can enhance one's own self-compassion. In this case, the chatbot, named Vincent was purposefully designed to share its foibles, such as embarrassingly admitting its mistakes, and seek advice from its user (see Figure 6.4, right). The case shows that a user can benefit from prosocial activities towards products, more so than when the products are designed to serve the users only.

FIGURE 6.4 Design cases that fall under the third opportunity: self as actor, others as receiver.

Sources: "Chooz" by Abby Lerner (left image), care-receiving Vincent. From Caring for Vincent: A Chatbot for Self-Compassion (Lee et al., 2019) (right image).

Opportunity 4: Others as Actor, Others as Receiver

The fourth opportunity is about facilitating activities in which the self (i.e., actor) is positively influenced by observing another person coming to the aid of someone else. Examples of this include feeling proud of a friend who volunteers at a homeless shelter, feeling inspired by a celebrity who uses their platform to raise awareness for a cause, or feeling touched by a stranger jumping into the water to save a child.

Elevation is the most frequently empirically studied reaction to watching someone help another person. Along with gratitude and admiration, elevation is one of three "other-praising" emotions (Tugade et al., 2014). Although all three involve recognizing someone else for a particular behavior, elevation is the result of witnessing virtuous activities or their moral excellence (Algoe & Haidt, 2009). It drives individuals to praise the person (and their deeds) and behave in a more prosocial way (Thomson & Siegel, 2013).

The Goedzak (Akkaya, 2009; meaning "Goodbag" in English) is a product that encourages acts of altruism within a community. The transparent bag, adorned with a vibrant yellow stripe, offers a second chance for people's old belongings (see Figure 6.5, left image). Many perfectly usable items are often discarded because it is inconvenient to take them to a thrift shop. By filling the bag with unwanted items and placing it on the street, people (i.e., others as actor) can easily create an opportunity for others (i.e., others as receiver) to freely collect valuable items instead of buying new ones. The Goedzak's attention-grabbing design ensures that these acts of generosity are visible to all, inspiring passers-by to appreciate the goodwill of Goedzak users (both donors and collectors) in giving old items a new

FIGURE 6.5 Design cases that fall under the fourth opportunity: Others as actor, others as receiver. Goedzak designed by Simon Akkaya (left image), rendition of Unsung Hero designed by Ogilvy & Mather Bangkok.

Sources: Studio HeijltjesAkkaya (left image); the image is the author's own rendition, created for illustrative purposes, based on the original content (right image).

purpose. Further, it raises awareness of the community's efforts towards sustainable consumption.

Another example is the commercial for Thai life insurance called "Unsung Hero" (Yahoo News, 2014). The three-minute video, designed by Ogilvy & Mather Bangkok, was created to elicit a feeling of elevation, stimulating prosocial contagion of cooperativeness (see Figure 6.5, right). It portrays a good Samaritan whose acts of kindness go initially unnoticed but, as the story unfolds, the viewers see the positive impact on those he helps, from helping a begging child get into school to reviving a dying plant. Viewers also witness displays of recipients' and onlookers' gratitude and a positive orientation towards him. The video shows the presence of a prosocial primary actor (i.e., others as actors) and eventual rewards for his community (i.e., others as receivers). Due to its effectiveness in evoking elevation in viewers, the video has been used in several emotion studies that focused on shared positive experiences (e.g., Sparks et al., 2019).

Towards Well-Being–Driven Design: Design Considerations

The four proposed design opportunities guide the design process, with an explicit focus on users' happiness. This requires designers to systematically envision how products, whether new or incrementally improved, will stimulate and support users and other relevant stakeholders to engage in positive activities. In the following sections, I discuss four challenges to consider when addressing positive activities: (1) dynamics of benefits of social activities, (2) ethics of designing for positive activities, (3) person-activity fit, and (4) assessment of the long-term well-being impact.

Dynamics of Benefits of Social Activities

Several empirical studies have demonstrated that people generally experience more pleasure and purpose from their activities when they engage others. In particular, the effects of positive activities within the third opportunity – the self (i.e., actor) and the others (i.e., recipient) – are strongly supported in the positive psychology literature because the positive activities within the category broadly meet all three basic psychological needs proposed by self-determination theory (Ryan & Deci, 2000): competence, autonomy, and relatedness. A note of caution, however, is that there are a few exceptions where people tend to be unhappier when they do an activity with someone else. For example, Dolan (2014) showed that commuting with someone else can be detrimental to happiness, perhaps due to the decreased level of autonomy, such as not being in control of deciding to take a preferred route or selecting the music. Similarly, doing homework can be much less meaningful when engaged in with someone else, especially when solitude allows people to be more productive.

Further, it is important to recognize that there is no one-to-one relationship between positivity and particular inter/intrapersonal activity. People may choose to be alone in doing something that they normally do with others when they are in a particular mood. As the well-being of an individual cannot be separated from their lived circumstances, assessing the contextual supports that shape experiences of well-being in both contexts (i.e., individual versus social) should be in the focus of attention. Therefore, designers need to be alert to these dynamics when envisioning usage contexts for their products, including who uses it, how it is used, how it is talked about, and what meanings to users (and others involved in the context) are allocated to it.

Ethics of Designing for Positive Activities

It is important to distinguish between personal values and moral values when determining what kinds of positive activities to design for. These two types of values do not always align. The values that guide one's positive activities can range from those that are morally acceptable to those that are morally unacceptable. To take an extreme example, playing a musical instrument at home can bring joy to the user because it aligns with their personal values of "stimulating vitality and developing talent," but the loud sound being generated may be intolerable to neighbors, especially at a late time. Aiming only to facilitate positive activities based on personal values, without holistically considering moral values, can potentially hinder the user's well-being or the well-being of others. Therefore, intentions to design for positive activities should be formulated with ethical sensitivity and an awareness of possible unwanted effects.

In the design process, the personal and moral values of users or other stakeholders may conflict, or they may be unaware of potential ethical consequences associated with using the product. It is, thus, necessary to examine the values that are affected by the product and ensure that they are morally acceptable. To achieve this, it is important to carry out ongoing and iterative critical examinations of how the use of the product can be seen as ethically legitimate. One strategy for this could be to include recurring cycles of ethical reflection in the design process (Roeser, 2010), in which the personal and moral values of users and other stakeholders are made explicit, foreseeing possible risks and what needs to be done to reconcile them from both individual and collective perspectives in both the short and long term.

Alternatively, raising awareness of differing perspectives on "what is ethical" can be the design goal of a product as often intended in the design approaches of critical design (Dunne & Raby, 2013) and reflective design (Sengers et al., 2005). These approaches use hypothetical products to challenge existing unquestioned social-cultural-technological norms; the products are meant to ask questions rather than solving problems. These approaches have been used to let people recognize ethical assumptions and dilemmas and provoke discussion with people with different perspectives in a variety of domains, including racial tolerance, politics, world hunger, cybersecurity, gender equality, health equity, and environmental sustainability, among others (for an overview of related design examples, see Ozkaramanli & Desmet, 2016). By iteratively confronting such products with the world, designers can gain new insights into diverse ethical perspectives on the topics of interest that, in turn, their next design iteration could embody.

Person-Activity Fit

Individuals play an active role in their own well-being (Layous & Lyubomirsky, 2013). Therefore, designing for well-being can have an intended impact only if users are actively involved in positive activities. When it comes to the effectiveness of positive activities, there will be differences caused by each person's preferences. It is important to consider that practicing positive activities requires more than knowing what to do, and engaging in intentional activities is not necessarily easy. For instance, developing a talent like learning a new language takes long-term, effortful practice. A person needs to transition from intention to action by remembering and overcoming obstacles in initiating them. This kind of self-regulatory effort demands considerable self-discipline and willpower (Lyubomirsky et al., 2005). Therefore, designers should be aware of and consider the characteristics of individuals or intended groups to motivate them to commit to particular positive activities.

One important factor is that every person has their unique character strengths, such as creativity, bravery, curiosity, forgiveness, appreciation of beauty, and prudence, among others (for an overview of the set of character strengths and virtues, see Peterson & Seligman, 2004). Character strengths are conceptualized as a "pre-existing capacity for a particular way of behaving, thinking, or feeling that is authentic and energizing to the user and enables optimal functioning, development, and performance" (Linley, 2008, p. 9). While all character strengths can contribute to one's well-being, some of them are a better personal fit than others. Research has shown a strong link between character strengths and well-being. When people are introduced to positive activities that align with their top character strengths, they feel most authentic, are intrinsically motivated, and experience higher fulfillment (Peterson & Seligman, 2004). Therefore, designers could benefit from investigating how person-activity fit can be supported by creating products that (1) reflect intended users' unique personal character strengths, (2) provide them with opportunities to increase awareness of their strengths, and (3) enable them to cultivate their strengths meaningfully. By catering to users' personal character strengths, products could be more enabling, motivating, and engaging to use, serving their purpose to foster particular positive activities. This would effectively support people to invest the time and energy necessary for long-term behavioral change and to effectively translate acquired knowledge into practice in a well-being–enhancing way.

Assessment of the Long-Term Well-Being Impact

Happiness denotes the extent to which people themselves subjectively believe or feel that their lives are going well. Consequently, it is often assessed based on self-reports (Nima et al., 2020). Measuring the long-term effects of products on user happiness poses several methodological challenges. First, evaluating products designed to foster well-being may take a considerable amount of time, unlike measuring moment-to-moment experiences such as emotional states. The intended effects could only become evident after a substantial period of product use. Second, becoming lastingly happy is not a linear process but rather involves setbacks in between. Therefore, longitudinal, repeated follow-up assessments are necessary to reliably capture the well-being effects. Existing approaches for assessing happiness effects, such as diaries, experience sampling method, ambulatory assessment, address some of these challenges, but an integrated approach has yet to be developed (for an overview of initial explorations, see Karahanoğlu & Ludden, 2021). Given these challenges, it is not surprising that longitudinal field studies remain rare in design research.

One alternative approach is to iteratively measure short-term predictors along with the long-term effects on well-being. For example, designers can explore whether a given feature (e.g., a reminder with a light effect on an interactive calendar) triggers users' interest in starting a positive activity (e.g., vividly replaying a vacation in one's mind). Determining whether products trigger or motivate the desired activity can be assessed in real-time, either during or shortly after the interaction, by utilizing data collection methodologies. For instance, an ecological momentary assessment (Shiffman et al., 2008) would allow intensive, repeated measurements of people's activities and constructs of happiness (e.g., psychological needs and mood) in real-world settings, and observing the patterns in their reports that unfold over time. Measuring short-term predictors would allow an understanding of how products are used, which activities they enable and for whom, and which barriers hinder adoption or engagement with them. Here, one issue is that although these approaches are increasingly used in both design research and psychology, they tend to require constant interruptions, making their applications too intrusive and increasing participants' fatigue in responding. This challenge has been prevalent in studies investigating design-mediated long-term subjective experiences (Schueller & Parks, 2014). Thus, further exploration on data collection tools and techniques that take these challenges into account when collecting nuanced subjective experiences should be undertaken.

Conclusion

Well-being theorists have generally been skeptical about the idea that owning and using products can be a reliable source of profound happiness. They often refer to research on materialism to support their critical perspectives. Materialism has been shown to be a strong predictor of unhappiness. People with high materialistic aspirations are less happy than those with low materialistic aspirations, even if they are able to fulfill these aspirations (Nickerson et al., 2003). To overcome this skepticism, design research has expanded its focus beyond products' material properties (e.g., appearance and material) to their impact on users' daily activities associated with well-being. This has resulted in a wealth of theoretical and empirical knowledge on the contribution of design to human happiness. In line with these efforts, this chapter aspired to advance our understanding of how designers can be supported in purposefully facilitating user happiness, with a focus on products as enablers of positive activities and users as the actor and recipient of such activities.

At a broad level, the proposed framework (see Table 6.1) provides designers with a conceptual structure and shared language to explore opportunities for enhancing positive experiences. It can stimulate rich discussions about key questions that the design team aims to address. Examples of such questions are: What positive activities can enhance users' and other stakeholders' happiness in a complementary manner? What daily contexts are associated with the positive activities of interest? What are the enablers and barriers to such positive activities for users and other stakeholders? The activity-focused approach embodied in the framework departs from many traditional design approaches that take the means (often a specific product type or technology) as the starting point for exploring and redirecting consequential effects. In contrast, the activity-focused approach considers users as actors and recipients, guiding strategic design decisions regarding positive activities to be embodied in human-product interactions.

Today's society demands design that prioritizes individual and societal values rather than just addressing economic and consumer demands and functional efficiency. Design is one of the most intimate and ubiquitous means of reaching people, representing a promising pathway for the human aspiration of living a good life. Products are an integral part of our daily lives, motivating, inspiring, and even challenging us. Depending on how they are designed, they can either enable or hinder us in our endeavors to foster happiness. Through a sociopsychological lens, this chapter delineated how products can make usage behaviors more meaningful with a focus on users' eudaimonic well-being and interpersonal peace, going beyond just providing instant feelings of stress-free relaxation and convenience. I hope this chapter provides inspiration and practical knowledge to design practitioners and students who want to systematically create positive user experiences.

Acknowledgments

This research was supported by the National Science Foundation (IIS-2143552) and the National Research Foundation of Korea (NRF-2020S1A5A2A03045893). I express my gratitude to the copyright owners of the photographs for their permission to use them in the manuscript.

References

Akkaya, H. (2009). *De Goedzak*. https://www.heijltjesakkaya.com/goedzak

Algoe, S. B., & Haidt, J. (2009). Witnessing excellence in action: The 'other-praising' emotions of elevation, gratitude, and admiration. *The Journal of Positive Psychology*, *4*(2), 105–127. https://doi.org/10.1080/17439760802650519

Anderson, C., Kraus, M. W., Galinsky, A. D., & Keltner, D. (2012). The local-ladder effect: Social status and subjective well-being. *Psychological Science*, *23*(7), 764–771. https://doi.org/10.1177/0956797611434537

Be My Eyes. (2023). *Describing Be My Eyes*. https://www.bemyeyes.com/press

Brickman, P. D., & Campbell, D. T. (1971). Hedonic relativism and planning the good society. In M. H. Appley (Ed.), *Adaptation level theory: A symposium*. Academic Press.

Calvo, R. A., & Peters, D. (2014). *Positive computing*. MIT Press.

Casais, M., Mugge, R., & Desmet, P. (2018). Objects with symbolic meaning: 16 directions to inspire design for well-being. *Journal of Design Research*, *16*(3–4), 247–281. https://doi.org/10.1504/jdr.2018.10020856

Cohen, S. (2004). Social relationships and health. *American Psychologist*, *59*(8), 676. https://doi.org/10.1037/0003-066x.59.8.676

Deci, E. L., & Ryan, R. M. (2008). Hedonia, Eudaimonia, and well-being: An introduction. *Journal of Happiness Studies*, *9*(1), 1–11. https://doi.org/10.1007/s10902-006-9018-1

Desmet, P. M. A. (2011, November 1). Design for happiness: Four ingredients for designing meaningful activities. In L. L. Chen, N. F. M. Roozenburg, & P. J. Stappers (Eds.), *Proceedings of the 4th world conference on design research* [CD-Rom]. TU Delft.

Desmet, P. M. A., & Fokkinga, S. (2020). Beyond Maslow's pyramid: Introducing a typology of thirteen fundamental needs for human-centered design. *Multimodal Technologies and Interaction*, *4*(3), 38. https://doi.org/10.3390/mti4030038

Desmet, P. M. A., & Pohlmeyer, A. E. (2013). Positive design: An introduction to design for subjective well-being. *International Journal of Design*, *7*(3), 1–15.

Diener, E., & Seligman, M. E. (2002). Very happy people. *Psychological Science*, *13*(1), 81–84. https://doi.org/10.1111/1467-9280.00415

Diener, E., & Tov, W. (2007). Subjective well-being and peace. *Journal of Social Issues*, *63*(2), 421–440. https://doi.org/10.1111/j.1540-4560.2007.00517.x

Dolan, P. (2014). *Happiness by design: Finding pleasure and purpose in everyday life*. Penguin.

Dourish, P. (2001). *Where the action is*. MIT Press.

Dunne, A., & Raby, F. (2013). *Speculative everything: Design, fiction, and social dreaming*. The MIT Press.

Eid, M., & Larsen, R. J. (2008). *The science of subjective well-being*. Guilford Press.

Erlhoff, M., & Marshall, T. (2008). *Design dictionary: Perspectives on design terminology*. De Gruyter.

Faulk, J., & Yoon, J. (2023). Does practicing diverse savouring techniques enhance subjective well-being? A randomised controlled trial of design-mediated savouring. *Journal of Design Research*, 1–29. https://doi.org/10.1504/jdr.2023.10060446

Grant, A. M., & Gino, F. (2010). A little thanks goes a long way: Explaining why gratitude expressions motivate prosocial behavior. *Journal of Personality and Social Psychology, 98*(6), 946. https://doi.org/10.1037/a0017935

Hassenzahl, M. (2010). *Experience design: Technology for all the right reasons.* Morgan & Claypool.

Hassenzahl, M., Burmester, M., & Koller, F. (2021). User experience is all there is. *I-com, 20*(3), 197–213. https://doi.org/10.1515/icom-2021-0034

Huber, M., Knottnerus, J. A., Green, L., Van Der Horst, H., Jadad, A. R., Kromhout, D., Leonard, B., Lorig, K., Loureiro, M. I., & Van der Meer, J. W. (2011). How should we define health? *BMJ, 343.* https://doi.org/10.1136/bmj.d4163

Huxley, A. (1950). *Themes and variations.* Harper.

Karahanoğlu, A., & Ludden, G. (2021). Imagining the future of longitudinal HCI Studies: Sensor-embedded everyday objects as subjective data collection tools. In *Advances in longitudinal HCI research* (pp. 101–120). Springer.

Koskinen, I. (2003). Empathic design in methodic terms. In I. Koskinen, K. Battarbee, & T. Mattelmäki (Eds.), *Empathic design. User experience for product design* (pp. 59–68). IT Press.

Kumar, A., Killingsworth, M. A., & Gilovich, T. (2014). Waiting for merlot: Anticipatory consumption of experiential and material purchases. *Psychological Science, 25*(10), 1924–1931. https://doi.org/10.1177/0956797614546556

Layous, K., & Lyubomirsky, S. (2014). The how, why, what, when, and who of happiness: Mechanisms underlying the success of positive activity interventions. In J. Gruber & J. Moscowitz (Eds.), *Positive emotion: Integrating the light sides and dark sides* (pp. 473–495). http://doi.org/10.1093/acprof:oso/9780199926725.003.0025

Layous, K., Nelson, S. K., Oberle, E., Schonert-Reichl, K. A., & Lyubomirsky, S. (2012). Kindness counts: Prompting prosocial behavior in preadolescents boosts peer acceptance and well-being. *PLoS One, 7*(12), e51380. https://doi.org/10.1371/journal.pone.0051380

Lee, M., Ackermans, S., Van As, N., Chang, H., Lucas, E., & IJsselsteijn, W. (2019, May). Caring for Vincent: A chatbot for self-compassion. In *Proceedings of the 2019 CHI conference on human factors in computing systems* (pp. 1–13). https://doi.org/10.1145/3290605.3300932

Linley, A. (2008). *Average to A+: Realising strengths in yourself and others.* ICON Group International.

Lyubomirsky, S. (2008). *The how of happiness.* Penguin Press.

Lyubomirsky, S. (2011). Hedonic adaptation to positive and negative experiences. In S. Folkman (Ed.), *The Oxford handbook of stress, health, and coping* (pp. 200–224). Oxford University Press.

Lyubomirsky, S., Sheldon, K. M., & Schkade, D. (2005). Pursuing happiness: The architecture of sustainable change. *Review of General Psychology, 9*(2), 111–131. https://doi.org/10.1037/1089-2680.9.2.111

Maslow, A. H. (1943). A theory of human motivation. *Psychological Review, 50*, 370–396.

Mogilner, C., Kamvar, S. D., & Aaker, J. (2011). The shifting meaning of happiness. *Social Psychological and Personality Science, 2*(4), 395–402. https://doi.org/10.1177/1948550610393987

Neff, K. D., Rude, S. S., & Kirkpatrick, K. L. (2007). An examination of self-compassion in relation to positive psychological functioning and personality traits. *Journal of Research in Personality, 41*(4), 908–916. https://doi.org/10.1016/j.jrp.2006.08.002

Nickerson, C., Schwarz, N., Diener, E., & Kahneman, D. (2003). Zeroing in on the dark side of the American dream: A closer look at the negative consequences of the goal for financial success. *Psychological Science*, *14*(6), 531–536. https://doi.org/10.1046/j.0956-7976.2003.psci_1461.x

Nima, A. A., Cloninger, K. M., Persson, B. N., Sikström, S., & Garcia, D. (2020). Validation of subjective well-being measures using item response theory. *Frontiers in Psychology*, *10*, 3036. https://doi.org/10.3389/fpsyg.2019.03036

Ozkaramanli, D., & Desmet, P. M. A. (2016, June 27). *Provocative design for unprovocative designers: Strategies for triggering personal dilemmas*. Design Research Society.

Peterson, C., & Seligman, M. E. (2004). *Character strengths and virtues: A handbook and classification* (Vol. 1). Oxford University Press.

Revord, J., Walsh, L. C., & Lyubomirsky, S. (2017). Positive activity interventions to enhance well-being: Looking through a social psychological lens. In *Subjective well-being and life satisfaction* (pp. 451–471). Routledge.

Roeser, S. (2010). Emotional engineers: Toward morally responsible design. *Science and Engineering Ethics*, *18*(1), 103–115. https://doi.org/10.1007/s11948-010-9236-0

Ryan, R. M., & Deci, E. L. (2000). Self-determination theory and the facilitation of intrinsic motivation, social development, and well-being. *American Psychologist*, *55*(1), 68–78. https://doi.org/10.1037//0003-066x.55.1.68

Ryan, R. M., & Deci, E. L. (2001). On happiness and human potentials: A review of research on hedonic and Eudaimonic well-being. *Annual Review of Psychology*, *52*(1), 141–166. https://doi.org/10.1146/annurev.psych.52.1.141

Schueller, S. M., & Parks, A. C. (2014). The science of self-help: Translating positive psychology research into increased individual happiness. *European Psychologist*, *19*(2), 145–155. https://doi.org/10.1027/1016-9040/a000181

Seligman, M. E. (2008). Positive health. *Applied psychology*, *57*, 3–18. https://doi.org/10.1111/j.1464-0597.2008.00351.x

Sengers, P., Boehner, K., David, S., & Kaye, J. J. (2005, August). Reflective design. In *Proceedings of the 4th decennial conference on critical computing: Between sense and sensibility* (pp. 49–58). http://doi.org/10.1145/1094562.1094569

Sheldon, K. M., & Lyubomirsky, S. (2006). Achieving sustainable gains in happiness: Change your actions, not your circumstances. *Journal of Happiness Studies*, *7*, 55–86. https://doi.org/10.1007/s10902-005-0868-8

Sheldon, K. M., & Lyubomirsky, S. (2012). The challenge of staying happier: Testing the hedonic adaptation prevention model. *Personality and Social Psychology Bulletin*, *38*(5), 670–680. https://doi.org/10.1177/0146167212436400

Shiffman, S., Stone, A. A., & Hufford, M. R. (2008). Ecological momentary assessment. *Annual. Review of Clinical Psychology*, *4*, 1–32. https://doi.org/10.1146/annurev.clinpsy.3.022806.091415

Slovak, P., Antle, A., Theofanopoulou, N., Daudén Roquet, C., Gross, J., & Isbister, K. (2023). Designing for emotion regulation interventions: An agenda for HCI theory and research. *ACM Transactions on Computer-Human Interaction*, *30*(1), 1–51. https://doi.org/10.1145/3569898

Sparks, A. M., Fessler, D. M., & Holbrook, C. (2019). Elevation, an emotion for prosocial contagion, is experienced more strongly by those with greater expectations of the cooperativeness of others. *PLoS One*, *14*(12), e0226071. https://doi.org/10.1371/journal.pone.0226071

Stephanidis, C., Salvendy, G., Antona, M., Chen, J. Y. C., Dong, J., Duffy, V. G., Fang, X., Fidopiastis, C., Fragomeni, G., & Fu, L. P. (2019). Seven HCI grand challenges. *International Journal of Human-Computer Interaction*, *35*(14), 1229–1269. https://doi.org/10.1080/10447318.2019.1619259

Tay, L., & Diener, E. (2011). Needs and subjective well-being around the world. *Journal of Personality and Social Psychology*, *101*(2), 354–365. https://doi.org/10.1037/a0023779

Thomson, A. L., & Siegel, J. T. (2013). A moral act, elevation, and prosocial behavior: Moderators of morality. *The Journal of Positive Psychology*, *8*(1), 50–64. https://doi.org/10.1080/17439760.2012.754926

Tugade, M. M., Shiota, M. N., & Kirby, L. D. (Eds.). (2014). *Handbook of positive emotions*. The Guilford Press.

United Nations. (2016). *The sustainable development goals report*. https://unstats.un.org/sdgs/report/2016/the%20sustainable%20development%20goals%20report%202016.pdf

Van Boven, L., & Gilovich, T. (2003). To do or to have? That is the question. *Journal of Personality and Social Psychology*, *85*(6), 1193–1202. https://doi.org/10.1037/0022-3514.85.6.1193

Veenhoven, R. (2011). Greater happiness for a greater number: Is that possible? If so, how. In K. M. Sheldon, T. B. Kashdan, & M. F. Steger (Eds.), *Designing positive psychology: Taking stock and moving forward* (pp. 396–409). Oxford University Press.

Verbeek, P.-P. (2011). *Moralizing technology: Understanding and designing the morality of things*. University of Chicago press.

Wahba, M. A., & Bridwell, L. G. (1976). Maslow reconsidered: A review of research on the need hierarchy theory. *Organizational Behavior and Human Performance*, *15*(2), 212–240. https://doi.org/10.1016/0030-5073(76)90038-6

Wheeler, J. A., Gorey, K. M., & Greenblatt, B. (1998). The beneficial effects of volunteering for older volunteers and the people they serve: A meta-analysis. *The International Journal of Aging and Human Development*, *47*(1), 69–79. https://doi.org/10.2190/vump-xcmf-fqyu-v0jh

Wicker, F. W., Brown, G., Wiehe, J. A., Hagen, A. S., & Reed, J. L. (1993). On reconsidering Maslow: An examination of the deprivation/domination proposition. *Journal of Research in Personality*, *27*(2), 118–133. https://doi.org/10.1006/jrpe.1993.1008

Yahoo News. (2014). *Tearjerking Thai life insurance ad: Try not to cry*. http://news.yahoo.com/thai-life-ad-unsung-hero-video-154250674.html?

Yoon, J., & Kim, C. (2022). Positive emodiversity in everyday human-technology interactions and users' subjective well-being. *International Journal of Human-Computer Interaction*, *14*, 1–16. https://doi.org/10.1080/10447318.2022.2121564

Yoon, J., Li, S., & Hao, Y. (2021). Design-mediated positive emotion regulation: The development of an interactive device that supports daily practice of positive mental time traveling. *International Journal of Human-Computer Interaction*, *38*(5), 432–446. https://doi.org/10.1080/10447318.2021.1948685

7

EMPOWERMENT BY DESIGN

Renata M. Leitão

What is the meaning of peace? Absence of war and struggle? For Indigenous and local communities who deal with the long-lasting legacy of colonialism, life includes a continuous struggle to defend not only their lands and territories but also their ways of being in the world (Escobar, 2020). Peace, the end of such struggles, can only be achieved through self-determination – that is, the right to live the kind of life they value (Salazar, 2009). Empowerment can be understood as the process of strengthening the capacity of marginalized people to shape their own lives.

In a time when Indigenous and marginalized individuals and communities throughout the globe are striving for their self-determination and redefining their place in the contemporary world, striving to forge a better life, this chapter argues that the practice of design can be a means to support their empowerment. It aims to offer a theoretical framework for understanding how design practice might contribute to the empowerment of Indigenous and marginalized individuals and communities by enhancing their power to act upon and ultimately change the material circumstances in which they live.

There are, however, a myriad of different definitions of 'design' and 'empowerment' that have little relation to the self-determination of oppressed communities. In this chapter, these two concepts – design and empowerment – refer to a fundamental ability of human beings: our ability to create and change the material conditions in which we live (Leitão & Marchand, 2018). The Brazilian educator Paulo Freire (1970) argued that to exist humanly is to be able to change the world. Design is one of the activities that allows us to transform our world in purposeful ways.

> As human beings, we continuously create things that help reshape reality and the essence of the world as we know it. When we create new

DOI: 10.4324/9781003462019-17

> things – technologies, organizations, processes, environments, ways of thinking, or systems – we engage in design. . . . To come up with an idea of what we think would be an ideal addition to the world, and to give real existence – form, structure, and shape – to that idea, is at the core of design as a human activity.
>
> *(Nelson & Stolterman, 2012, p. 1)*

Paulo Freire probably was not aware of the term 'design' in 1968, when he first published *The Pedagogy of the Oppressed*, but he was surely referring to what has been conceptualized as 'design' when he wrote:

> It is as transforming and creative beings that humans, in their permanent relations with reality, produce not only material goods – tangible objects – but also social institutions, ideas, and concepts. Through their continuing praxis, men and women simultaneously create history and become historical-social beings.
>
> *(1970, p. 101)*

We continuously create artificial things – instruments, symbols, artifacts, buildings, policies, institutions, environments, services, and so on – in striving to flourish and achieve that which we imagine to be a good life (Leitão & Roth, 2020). And so, *design is the form of intelligence that we engage to bring the intangible (ideals, desires, aspirations) into the tangible.*

In the pursuit of the good life (as each culture defines it), people and communities evolve and interact with their environment and circumstances in creative ways. No culture has ever been static. UNESCO suggests we understand cultures as "systems that continually evolve through internal processes and in contact with the environment and other cultures" (2009, p. 9) – in other words, cultures can be regarded as trajectories toward the future. Therefore, cultural diversity represents numerous creative trajectories.

Colonialism, imperialism, and globalization have spread throughout the planet. The cultural template originating in Europe, known as modernity, imposed the belief that only one trajectory is possible or desirable: the one that follows the project of modernity. This belief has generated many forms of oppression, violence, and forced assimilation that ruined the autonomous trajectories of many Indigenous and non-Western cultures. Before European colonization, Indigenous peoples lived in a world of their own making. The term Indigenous comprises the long-standing inhabitants of any part of the world before European colonization. This paper refers to numerous distinct populations, who live in different contexts, with distinct cultures, lifestyles, and experiences. Despite these vast differences, there are two key similarities between Indigenous peoples: they share the legacy of colonization and the struggle for self-determination.

In the last few centuries, numerous populations were forced into the Eurocentric model of social and material life. The struggle for self-determination represents the assertion of Indigenous peoples' right to conceive well-being and the Good Life in their own way, on their own terms, instead of conforming to the model of the Global North.

And here lies the importance of empowerment. Empowerment can be broadly defined as "a progression that helps people gain control over their own lives and increases the capacity of people to act on issues that they themselves define as important" (Luttrell & Quiroz, 2009, p. 19). Empowerment can be understood as a process of strengthening the capacity of marginalized people to shape their own destiny in contexts where power asymmetries have hindered it (Bartlett, 2004).

Since the late 1980s, the term 'empowerment' has been widely used in the rhetoric of the fight for social justice. Nonetheless, the term has received a myriad of different definitions and contradictory meanings. For example, some people understand empowerment as a political concept involving the collective struggle of the poor and oppressed; for others, it alludes to autonomy, self-confidence, self-worth, and mobilization (Jupp & Ali, 2010; Luttrell & Quiroz, 2009), while others believe empowerment is about access to information and technologies (Janzer & Weinstein, 2014). The goal of this chapter is not to present the multitude of different interpretations and definitions of the term 'empowerment.' The goal is to present the perspective and theoretical framework I have adopted as a social designer – while acknowledging that other perspectives and frameworks might also be valid.

Context

I became a graphic designer in the early 2000s, when bold claims about the power of design to change the world were made. A new field of design focused on social impact was emerging: social design. There are several terms that are used to refer to the field of social design: design for social innovation, design for social change, design for social need, human-centered design, and so on (DiSalvo et al., 2011; Janzer & Weinstein, 2014). Introduced to social design, I learned that design could address intractable social problems, contribute to social change, improve lives, and empower underserved communities. Social design methods, such as Double Diamond or IDEO's Human-Centered Design, became widely popular. Many designers wished to do good and promote social justice, instead of catering to the market, including me.

An example of the optimism with designing for 'underserved' communities is the book *Design for the Other 90%* by Cynthia Smith, published in 2007 in conjunction with a major exhibition at the Smithsonian's Cooper-Hewitt, National Design Museum. The cover features a product, the LifeStraw, designed as a solution for those who do not have access to safe drinking water (see Figure 7.1). In 2011, Smith published another book: *Design with the Other 90%: Cities*, which changed the preposition from *for* to *with*.

In spite of (or because of) the bold claims, several local communities and designers from the Global South have criticized social design initiatives as white saviorism, neo-colonialism, or imperialism. As much as I believe in the positive force of design, I also recognize the danger of well-intentioned but ill-informed interventions. When I started to work with Indigenous people, I had the same goal as most social designers: to design a solution to their problems. With time, I realized that this 'solution' was not nearly as important as the capacity of local stakeholders to define and act upon their own projects.

My Trajectory

In 2003, out of design school, as a graphic designer, I intended to work on promoting awareness about important issues. I worked for five years on the

FIGURE 7.1 An example of the optimism with designing for "underserved" communities.

Source: Design for the Other 90% by Cynthia E. Smith © Cooper Hewitt, Smithsonian Design Museum, 2007. Image: © 2005 Vestergaard Frandsen.

communication of corporate social responsibility (CSR) and campaigns for social causes, until I had a professional crisis. I realized that it is very difficult to separate the communication of initiatives for good causes from communication that stimulates consumerism. Greenwashing (and its sibling, socialwashing) runs rampant. The problem with greenwashing and socialwashing is not only the false claims but their power to turn transformative concepts – such as sustainability, empowerment, and decoloniality – into empty clichés and commodify them. A 'good-cause' industry has been created to nurture ego-driven desires and delusions of 'saving the world' with events, magazines, consultants, and so on.

For example, *Viagem de Canoa* (Canoe Trip) is an initiative created by the native Caiçara communities of Guaraqueçaba (Brazil) and led by Renato Siqueira to take control of the narratives connected to their territory and livelihoods. Guaraqueçaba is home to a 300,000-hectares forest conservation area and received one of the world's pioneering carbon offset projects funded by three North American corporations – General Motors, American Electric Power, and Chevron. Unfortunately, this project, environmental policies and conservation activities destroyed the livelihoods of local Caiçara communities (Terra de Direitos, 2015; World Rainforest Movement, 2011). And yet, for many years, it was very difficult to have their struggles acknowledged by outsiders, as the environmentalists were often depicted as the "good guys" in a quest to save the planet. In response to the dire circumstances, Caiçaras have invited researchers, students, journalists, and filmmakers to make a canoe trip through their territory and learn their perspectives and ways of knowing. Therefore, for over 15 years, they have educated potential allies and raised awareness of their struggles and conflicts with environmentalists.

In my graphic design practice, communicating CSR, I frequently used images of Indigenous or Black individuals and communities to portray "the-poor-other-in-need." Those people were confined to the role of *the needy* who received the help of virtuous entities (a.k.a. white saviors). I started to wonder if the initiatives I was communicating really benefited the communities they were portraying or if their photos were mostly benefitting the public image of the donor. As I worked with photos of poor communities almost every day, several questions came to my mind: Who are those people? What do they have to say? How do they want to be seen? What kind of project can break the cycle of poverty and really benefit them? In 2009, those questions led me back to school to become a social designer.

Can the Wish to Do Good Really Do Good?

This question has guided my academic trajectory. At issue here are not goals and intentions – most designers agree on the goals of promoting social justice and environmental sustainability. At issue here are the means and the specific outcomes we aim to create in our pursuit of social justice. Doing good can take many different forms. We can act with conviction that we are doing good for a community, and our actions end up being harmful to the people we are trying to help. Unquestionably, there is a vast difference between designing products, buildings, and visual identities (the traditional forms of design) and designing social change in a society we barely know. The risks of well-intended but harmful actions in intercultural initiatives are very real.

While social design initiatives had been previously criticized by designers from the Global South (Tunstall, 2013), Bruce Nussbaum stirred controversy in 2010 with his article "Is Humanitarian Design the New Imperialism?" He asked: "Are designers the new anthropologists or missionaries, come to poke into village life, 'understand' it and make it better – their 'modern' way?" (Nussbaum, 2010, p. 1), wondering if designers' desire to help has done more harm than good. Nussbaum particularly mentioned projects focused on technology-driven transformation in which outsiders design a product to solve a problem of the receiving communities, using a top-down approach. Perhaps the most emblematic example of the failure of such an approach to generate social change is One Laptop Per Child (OLPC), developed by the MIT Media Lab. OLPC aimed to distribute low-cost laptop computers to empower the world's poorest children through education (Negroponte, 2005). There are several papers and books that examine the shortcomings of the OLPC as a cautionary tale that social change, empowerment, or bridging the digital divide cannot be accomplished by merely providing people with computers (Ames, 2019; Fajebe et al., 2013; Janzer & Weinstein, 2014; Srinivasan, 2017; Yujuico & Gelb, 2011).

There are multiple cases of techno-solutions that fail to create meaningful change (Campbell, 2017; Martin et al., 2023), even though design researchers are not encouraged by funding structures to share our failures. Researchers are rarely funded to study the failures of other projects so the field can learn from our shortcomings and evolve. Moreover, the communities that receive the techno-solutions do not participate in the academic production of knowledge – their voices are seldom heard.

Designing solutions for or with poor communities, even with the best intentions, can be disempowering, as it fixates the distinction between the people with problems (communities) and the people with solutions (designers) – disempowering because it makes local people believe that only outsiders (who received education in the Eurocentric model) have the power to produce any change (Tuck, 2009).

Disempowerment by Design

The bias of dismissing the creativity, innovation, and intelligence of Indigenous individuals and communities is a fundamental element of the power asymmetries established by colonialism/imperialism. Since the 15th century, Indigenous and colonized societies (which today form the Global South) have been portrayed as inferior – ignorant, savage, backward. In other words, belonging to a culture that has been colonized by Europeans has been equated with a form of disability. Māori scholar Linda Tuhiwai Smith explained how imperialism has framed the lives of Indigenous peoples worldwide:

> One of the supposed characteristics of primitive peoples was that we could not use our minds or intellects. We could not invent things, we could not create institutions or history, we could not imagine, we could not produce anything of value, we did not know how to use land and other resources from the natural world, we did not practice the 'arts' of civilization. By lacking such virtues we disqualified ourselves, not just from civilization but from humanity itself. In other words we were not 'fully human'; some of us were not even considered partially human.
>
> *(Smith, 1999, p. 20)*

Arguably, attempts to 'empower' Indigenous people by providing them with Western tools, knowledge, and technologies can be compared to attempts to extinguish a fire using gasoline. In moments of crisis, such as after a natural disaster, direct assistance (food, medication, tools, and material goods) from outsiders is fundamental. However, empowerment is important in the context of chronic circumstances that prevent individuals and collectivities from being able to shape their lives on their own terms (Leitão, 2022). Frequently these circumstances materialize in the form of poverty. Arjun Appadurai (2004) states that poverty is inequality (in the distribution of power) materialized. Asymmetries of power are fortified when one side is always the powerful one who conceives all the possible solutions and the other is the ever-needy in a subordinate role.

I use the expression *Prometheus complex* (Leitão & Noel, 2022) to describe the tendency of designers to imagine they are the givers of enlightening fire to poor and underserved peoples. There is nothing empowering about this mindset. On the contrary, it undermines the creative power of the communities we design with or for, which is in itself a form of oppression (Freire, 1970).

> Contrary to the widespread assumption that abundant resources foster innovation, we believe that, frequently, a lack of resources triggers creativity. Because we came from the Global South, we witnessed that local innovation is often stifled by the influence of external experts that undermines the creative confidence of local (lay) designers.
>
> *(Leitão & Noel, 2022, pp. 7–8)*

Empowering Design

We have seen a shift from the first form of social design – designing *for* the other (the poor and underserved) – to designing *with* the other, which uses human-centered design approaches to get communities' input. At this moment, as suggested by Manzini (2015), I believe designers need to take another step: support the ability of people to design for themselves. This stance entails a redefinition of the role of the social designer: someone who supports with their knowledge and skills the talents of people to define and act upon their own projects (Manzini, 2015). The ultimate goal is therefore to increase the stakeholders' autonomy to a point where they become independent of the designers. This is why these stakeholders' empowerment is so crucial to the success of such collaborations.

Several design scholars (e.g., Cross, 2006; Fry, 2009; Manzini, 2015; Nelson & Stolterman, 2012) claim that design is an innate talent, ability, or form of intelligence that all human beings possess. However, the way this ability is expressed has varied considerably across different epochs, societies, and cultures (Otto & Smith, 2013). As with any human talent, it must be cultivated through practice (Fry, 2009; Manzini, 2015). And what does it mean to acquire design practice? It means learning how to "think and act by projects" (Manzini, 2015, p. 68). This chapter argues that the capacity to create and realize projects might be an important lever of self-determination and empowerment for local communities striving to redefine their place in the world and forge a better life. Designers can support the empowerment of members of Indigenous communities by nurturing their inherent capacity to design.

This nurturing may take the form of recognizing grassroots innovation and supporting the work of local lay designers (i.e., creative individuals who have not received Eurocentric design education) to address the problems facing their communities.

> Although human-centered design is participatory, the focus is on understanding users and their context rather than exploring how people are already designing to meet their own needs. Social innovation is shedding some light on how lay designers are designing in their contexts, but the majority of professional designers have not been trained to consciously seek and identify lay designers within communities. This oversight has resulted in professional designers often missing the valuable insights and, in some cases, solutions offered by lay designers.
>
> *(Campbell, 2017, p. 31)*

Collaborations with local lay designers might also be empowering for other community members as the value of their agency and innovative capacity is recognized. Workshops that join local creatives and professional designers might also foster communities' assets and their talent for design. Unfortunately, most designers have not been trained to recognize the innovative capacity, creativity, and intelligence of local stakeholders – on the contrary, designers are trained to focus on their needs and problems, creating the disempowering power dynamic in which one group always has problems and the other solutions. The framework I describe in the following is a contribution to a better understanding of what might be an empowering design.

Empowerment by Design

This chapter adopts the definition proposed by social economist Naila Kabeer: "empowerment . . . refers to the expansion in people's ability to make strategic life choices in a context where this ability was previously denied to them" (2001, p. 19).

Kabeer understands strategic life choices as critical for people to live the kind of lives they want, determining the parameters of such lives. For her, the process of empowerment is inescapably tied to the condition of disempowerment. "People who exercise a great deal of choice in their lives may be very powerful, but they are not empowered in the sense in which I am using the word, because they were never disempowered in the first place" (Kabeer, 2001, p. 19). Kabeer's work focuses on women's empowerment in South Asia, but the expression "a context where this ability was denied to them" can perfectly refer to the consequences of assimilation and development policies on numerous Indigenous and local communities.

Indigenous peoples used to live in a world of their own making and have been forced into the Eurocentric model of social and material life. And so, their material life has been dramatically altered over the last few centuries. Several Indigenous peoples are surrounded by immediate material circumstances that they did not choose. For instance, the Caiçaras (in Guaraqueçaba, Brazil) were suddenly thrown into the money economy in the 1990s, when many of their subsistence practices were forbidden (Leitão et al., 2013). Consequently, they were forced to restructure their livelihoods in order to buy food and industrialized goods from outside. Consuming industrialized products also increased the amount of garbage in their territory and changed the meaning and function of their craft traditions. Similarly, the Atikamekw (in Quebec, Canada) were originally semi-nomadic hunters and gatherers (Poirier, 2004). They had to completely restructure their ways of life since the 1950s when they started to live on a permanent basis on three small reservations, which represent small fractions of their ancestral territory. Nowadays many Atikamekw live in overcrowded houses that are unfit for their lifestyle. Both populations must deal with the material wreckage brought about by assimilation policies.

I believe strategic life choices are necessarily linked to a project: a life project, a community project, an identity project, or a societal project. The life one wants must be envisioned in order to guide one's choices. If people do not have a project of a desirable situation, no choice can be "strategic" in Kabeer's sense. The ability to make strategic life choices can be enhanced by the practice of design.

Qualities of Choices

Kabeer (2012) states that not every choice is relevant to a process of empowerment. (See Figure 7.2). She described three qualities for their relevance:

(1) **The possibility of choosing otherwise.** Meaningful choices are made with real alternatives, and they differ from choices made out of the fear of negative consequences or absence of alternatives.
(2) **The perception of alternatives.** Meaningful choices also entail providing awareness of possible alternative courses of action. Often disempowered people are not even able to consider that there can be an alternative to the situation they are in (Luttrell & Quiroz, 2009).

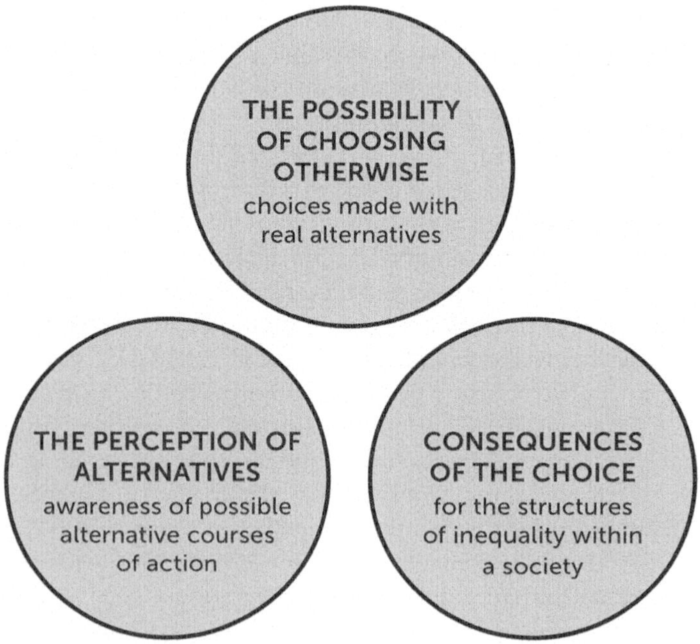

FIGURE 7.2 Qualities that make a choice relevant to a process of empowerment described by Kabeer (2012).

(3) **The consequences of choice for the structures of inequality within a society.** Meaningful choices seek to transform social structures and not merely reproduce them.

While the possibility of choosing otherwise and the consequences of choice for the structures of inequality within a society are essential for any empowerment process, this chapter will focus on the second quality – the perception of alternatives – because it can be particularly fortified by the practice of design.

Perceived Alternatives and Design

How can people's capacity to make projects be rekindled? As already stated, disempowerment sometimes prevents people from even considering there can be an alternative to the situation they are in (Luttrell & Quiroz, 2009). We suggest that the first step toward empowerment is nurturing – and creating spaces to nurture – their imagination.

> The concept of imagination can be used to describe something that is not given, per se, but that exists as possibility, and that nevertheless still has a present effect as, created in the minds of people, it guides their behavior, ideas, and orientation in the world.
>
> *(Folkmann, 2014, p. 7)*

Envisioning that which does not (yet) exist enables us to strive to bring it to existence. For Folkmann (2014), imagination is related to the enabling and stimulation of possibilities. The ability to create images of desirable situations is likely to enable individuals to envision new possibilities. We can nurture stakeholders' imagination by enhancing their image-making skills (e.g., drawing, photography, participatory video, photovoice, prototyping) and using design methods. Visual methods transform abstract ideas into something visible and tangible enabling designers to share those ideas with others; to discuss them; to explore their feasibility, viability, and possible outcomes; to identify opportunities; and so on. In this sense, abilities related to image-making can fuel action-oriented imagination, and especially collective imagination.

Considering that a process of change usually unfolds amid several actors, the ability to communicate images of that-which-is-not-yet is important in the process of collective imagination. Visions of possible outcomes must be shared among stakeholders in order to catalyze community reflection, build consensus, and, finally, mobilize collective action. Could verbal discussions about stakeholders' visions, without the use of images, have the same impact on their empowerment? Srinivasan's research (2012) indicates the primacy of images, over verbal discussions, as a means to catalyze the emergence of community aspirations, along with consensus and collective action.

Every project has two stages: envisioning and realizing. Striving to bring the imagined outcome to reality mobilizes resources and concrete actions, setting in motion the process of intentional change.

Dimensions of Empowerment

Kabeer (2001) conceptualizes three interrelated dimensions of empowerment: (1) resources, (2) agency, and (3) achievements (see Figure 7.3). Changes in each dimension build on, contribute to, and benefit from changes in the others (Gammage et al., 2016).

(1) **Resources**: the conditions under which choices are made (Kabeer, 2001). They comprise material resources but also the social and human resources that could enhance the ability to exercise choice.

> Human resources are embodied in the individual and encompass his or her knowledge, skills, creativity, imagination and so on. Social resources, on the other hand, are made up of the claims, obligations and expectations which inhere in the relationships, networks and connections which prevail in different spheres of life and which enable people to improve their situation and life chances beyond what would be possible through their individual efforts alone.
>
> *(Kabeer, 2001, p. 20)*

For Kabeer, "the terms on which people gain access to resources are as important as the resources themselves when the issue of empowerment is being considered" (2001, p. 20). Access to resources may be provided through clientelist and disempowering relationships or may be achieved in ways that enhance dignity and a sense of self-worth. In this view, empowerment is not only about access to new resources but the recognition and cultivation of the resources one already possesses.

(2) **Agency**: the ability to define one's goals and act upon them (Kabeer, 2001). She explains that agency is much more than outward action, encompassing the meaning, motivation, and purpose that individuals bring to their activities. In short, agency can be understood as the capacity for purposive action (Gammage et al., 2016, p. 6). We can argue that design is itself a form of agency.

(3) **Achievements**: The outcomes of choices (Kabeer, 2001). To 'achieve' entails translating agency and resources into tangible outcomes. Kabeer argues that a fundamental manifestation of disempowerment is the inability to achieve. An empowered group has an increased capacity to convert choices, resources, and agency into tangible outcomes in comparison to a disempowered one.

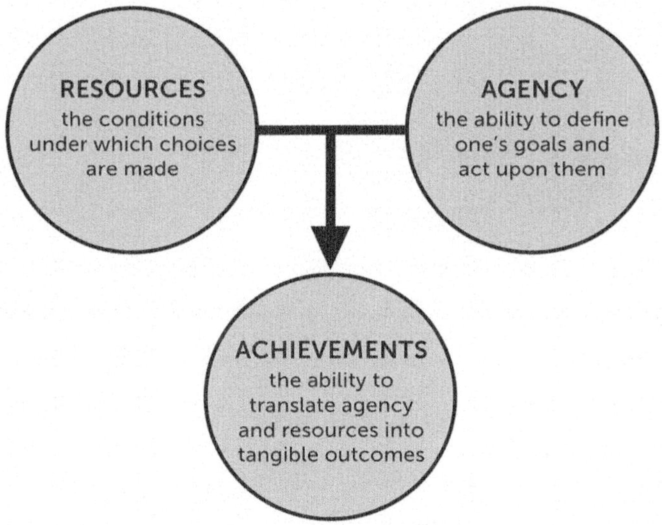

FIGURE 7.3 The three interrelated dimensions of empowerment conceptualized by Naila Kabeer (2001).

In this conception, empowerment entails developing the ability to purposefully reshape the material conditions in which we live, an ability that we can name design.

Empowerment and Design Training

This outlook of empowerment by design entails that, while we form collaborations to address local problems, designers train community members to develop their talent to design. How could design training given by Western designers be empowering to members of Indigenous communities? Our source of inspiration is Paulo Freire's emancipatory approach to adult literacy. Freire (1970) was critical of what he called banking education, wherein learners are mere receivers of "valid" knowledge from the teacher. Instead, Freire proposes a problem-posing education, conceived as a partnership between teacher and student to investigate themes that are meaningful to the student's everyday life. In this thematic investigation, students not only develop new skills and acquire knowledge, but they become aware of the circumstances that surround them and develop the capacity to express their ideas and to act upon the world in order to transform it (see Figure 7.4).

Design training in local communities can follow the model of a banking education when the teacher believes that design is a capacity that only design experts, trained in Eurocentric schools, possess (Filho, 2013). On the other hand, design educators can adopt a problem-posing approach, considering design as a natural

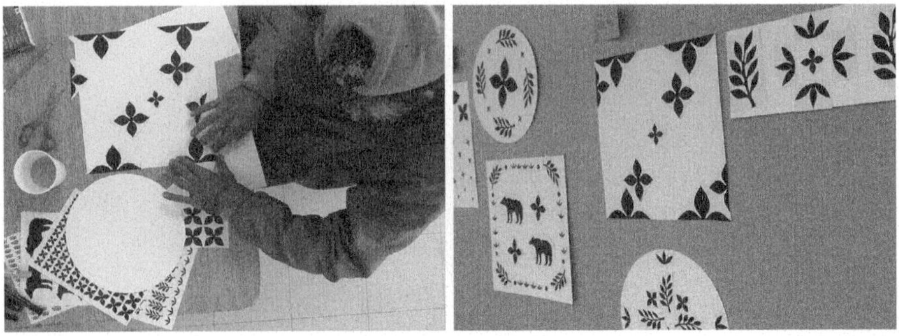

FIGURE 7.4 Graphic design training with Atikamekw participants, inspired by Paulo Freire's pedagogy. Participants created visual compositions using fundamental symbols of Atikamekw Nation's iconography (Leitão & Marchand, 2017).

FIGURE 7.5A AND 7.5B Graphic design training with Atikamekw participants inspired by Paulo Freire's pedagogy.

human ability that can be nurtured. Therefore, we move from providing students with what they lack to cultivating what they have (See Figure 7.5A & 7.5B).

We might enhance the participants' talent to design (and consequently their agency) by challenging them. The challenge consists of short-term design projects that have to be accomplished with limited time and resources. For Freire, a problem-posing approach must be challenging.

> Students, as they are increasingly posed with problems relating to themselves in the world and with the world, will feel increasingly challenged and obliged to respond to that challenge. . . . Their response to the challenge evokes new challenges, followed by new understandings; and gradually the students come to regard themselves as committed.
>
> *(1970, p. 81)*

The facilitators (expert designers) challenge the participants to explore the problem and, at the same time, discover how skillful, talented, and knowledgeable they can be. The use of limited external resources forces the participants to recognize the resources they already have at hand. Foreign elements can be gradually introduced to help the participants create the responses to the challenge. This approach aims to enhance their motivation and confidence in their ability to act (Leitão & Marchand, 2018).

Design and Achievements

I argue that design workshops held in disempowered Indigenous communities are essentially exercising the participants' capacity to achieve – converting choices, actions, and resources into tangible outcomes. Specific short-term design projects serve as a small-scale experience in which the participants establish a concrete goal, a project, and act to materialize it. In this process, they recognize their own resources: skills, creativity, cultural heritage, material and human resources available in the community. At the end of the workshops, they must produce tangible outputs. By observing their outputs and receiving feedback, participants can reflect on the results of their choices and actions and compare with their initial intention.

In this sense, we are not only metaphorically teaching participants how to fish. We will cook the fish together, taste the fish, and discuss what was successful or not: Was the cooking method appropriate? How was the seasoning? How could the process be improved? And then we can try again.

In my experience, I realized that small short-term projects have more impact on the participants' self-confidence than long-term overarching ones because the circuit of establishing a goal, realizing a project, reflecting upon the results, and identifying future opportunities is shorter. A sequence of small-scale projects becomes the basis for reflection and discussion about the future, with the ultimate goal of enhancing the participants' confidence in their ability to act and accomplish projects so that they can define and put into practice their own projects in their everyday lives (Leitão & Marchand, 2017, 2018) (see Figure 7.6). As part of my doctoral research, I led the development of the

FIGURES 7.6 Collection of textiles designed and printed by artists of the Atikamekw nation in collaboration with designers of the University of Montreal in the Tapiskwan workshops (Leitão & Marchand, 2017).

methodology of Tapiskwan workshops, a participatory action research project and collaboration between Indigenous Atikamekw artists, artisans, knowledge keepers, and youth with a design team of Université de Montréal, directed by Professor Anne Marchand.

Conclusion

In this chapter, we adopted a notion of empowerment related to the capacity to shape the material conditions in which one lives. An empowered individual or collectivity develops the capacity to make projects and plans, envisioning a future situation and mobilizing their resources to make their project happen. This form of empowerment enables individuals and communities to work with material circumstances as they are here and now – not waiting for eventual political changes from above. And since design is a practice of making and realizing projects, we believe that it can contribute to the empowerment of members of Indigenous communities by nurturing their capacity to imagine creative responses to challenges.

Acknowledgments

This work was supported by the Social Sciences and Humanities Research Council of Canada (SSHRC) under the Insight Development Grant.

References

Ames, M. G. (2019). *The charisma machine: The life, death, and legacy of one laptop per child*. MIT Press.

Appadurai, A. (2004). The capacity to aspire: Culture and the terms of recognition. In V. Rao & M. Walton (Eds.), *Culture and public action*. Stanford University Press.

Bartlett, A. (2004). *Entry points for empowerment*. CARE Bangladesh.

Campbell, A. D. (2017). Lay designers: Grassroots innovation for appropriate change. *Design Issues, 33*(1), 30–47. https://doi.org/10.1162/DESI_a_00424

Cross, N. (2006). *Designerly ways of knowing*. Springer-Verlag.

DiSalvo, C., Lodato, T., Fries, L., Schechter, B., & Barnwell, T. (2011). The collective articulation of issues as design practice. *CoDesign, 7*(3-4), 185–197. https://doi.org/10.1080/15710882.2011.630475

Escobar, A. (2020). *Pluriversal politics: The real and the possible*. Duke University Press.

Fajebe, A. A., Best, M. L., & Smyth, T. N. (2013). Is the one laptop per child enough? Viewpoints from classroom teachers in Rwanda. *Information Technologies & International Development, 9*(3), 29–42.

Filho, E. R. (2013). Design and craftsmanship: The Brazilian experience. *Design Issues, 29*(3), 64–74. https://doi.org/10.1162/desi_a_00221

Folkmann, M. N. (2014). Unknown positions of imagination in design. *Design Issues, 30*(4), 6–19. http://doi.org/10.1162/DESI_a_00293

Freire, P. (1970). *Pedagogy of the oppressed*. Continuum.

Fry, T. (2009). *Design futuring: Sustainability, ethics and new practice*. Berg.

Gammage, S., Kabeer, N., & van der Meulen Rodgers, Y. (2016). Voice and agency: Where are we now? *Feminist Economics, 22*(1), 1–29. https://doi.org/10.1080/13545701.2015.1101308

Janzer, C. L., & Weinstein, L. S. (2014). Social design and neocolonialism. *Design and Culture, 6*(3), 327–343. https://doi.org/10.2752/175613114x14105155617429

Jupp, D., & Ali, S. I. (2010). *Measuring empowerment? Ask them* (Sida Studies in Evaluation). Swedish International Development Cooperation Agency.

Kabeer, N. (2001). Reflections on the measurement of women's empowerment. In A. Sisask (Ed.), *Discussing women's empowerment—Theory and practice* (pp. 17–57). Sida.

Kabeer, N. (2012). Empowerment, citizenship and gender justice: A contribution to locally grounded theories of change in women's lives. *Ethics and Social Welfare, 6*(3), 216–232. https://doi.org/10.1080/17496535.2012.704055

Leitão, R. M. (2022). From needs to desire: Pluriversal design as a desire-based design. *Design and Culture, 14*(3), 255–276. https://doi.org/10.1080/17547075.2022.2103949

Leitão, R. M., & Marchand, A. (2017). Graphic design as an instrument of identity assertion for Indigenous peoples: The case of the Tapiskwan project. *Communication Design, 5*(1–2), 203–223. https://doi.org/10.1080/20557132.2017.1385252

Leitão, R. M., & Marchand, A. (2018). Le design et l'empowerment au sein des communautés autochtones: s'engager avec la matérialité. *Recherches amérindiennes au Québec, XLVIII*(1–2), 91–99. https://doi.org/10.7202/1053706ar

Leitão, R. M., Marchand, A., & De Coninck, P. (2013). Craftsmanship as a means of empowerment for the Caiçara population of Guaraqueçaba. *The International Journal of Design in Society, 6*(3), 97–108. https://doi.org/10.18848/2325-1328/cgp/v06i03/38509

Leitão, R. M., & Noel, L.-A. (2022). If you're not in an existential crisis as a designer in Social Design, you're not doing it right! In E. Tsekleves & L.-A. Noel (Eds.), *The little book of designer's existential crises in 2022* (pp. 5–11). Imagination Lancaster.

Leitão, R. M., & Roth, S. (2020). Understanding culture as a project: Designing for the future of an Indigenous community in Québec. *FormAkademisk, 13*(5).

Luttrell, C., & Quiroz, S. (2009). *Understanding and operationalising empowerment.* Overseas Development Institute.

Manzini, E. (2015). *Design when everybody designs: An introduction to design for social innovation.* MIT Press.

Martin, C., Cross, J., & Verhoeven, A. (2023). Displaced energy practices: Autonomous design in the context of humanitarian intervention in Goudoubo camp, Burkina Faso. In S. A. Lichtman & J. Traganou (Eds.), *Design, displacement, migration: Spatial and material histories.* Routledge.

Negroponte, N. (2005). *One laptop per child (OLPC).* Retrieved September 12, 2023, from https://ilp.mit.edu/node/37559

Nelson, H. G., & Stolterman, E. (2012). *The design way: Intentional change in an unpredictable world* (2nd ed.). The MIT Press.

Nussbaum, B. (2010). *Is humanitarian design the new imperialism? Fast Company's co.design.* http://www.fastcodesign.com/1661859/is-humanitarian-design-the-new-imperialism

Otto, T., & Smith, R. C. (2013). Design anthropology: A distinct style of knowing. In W. Gunn, T. Otto, & R. C. Smith (Eds.), *Design anthropology: Theory and practice* (pp. 1–29). Bloomsbury.

Poirier, S. (2004). The Atikamekw: Reflections on their changing world. In R. B. Morrison & C. R. Wilson (Eds.), *Native peoples: The Canadian experience* (pp. 129–150). Oxford University Press.

Salazar, J. F. (2009). Self-determination in practice: The critical making of Indigenous media. *Development in Practice, 19*(4–5), 504–513.

Smith, L. T. (1999). *Decolonizing methodologies: Research and Indigenous peoples.* Zed Books.

Srinivasan, R. (2012). Rethinking digital cultures and divides: The case for reflective media. *The Information Society, 28*(1), 24–36.

Srinivasan, R. (2017). *Whose global village? Rethinking how technology shapes our world.* New York University Press.

Terra de Direitos. (2015, June 30). Pescadores e pescadoras artesanais do Paraná derrotam Instituto Chico Mendes de Conservação e Biodiversidade – ICMBio na justiça. *Terra de Direitos.* Retrieved May 13, 2024, from https://terradedireitos.org.br/noticias/noticias/pescadores-e-pescadoras-artesanais-do-parana-derrotam-instituto-chico-mendes-de-conservacao-e-biodiversidade-icmbio-na-justica/18088

Tuck, E. (2009). Suspending damage: A letter to communities. *Harvard Educational Review, 79*(3), 409–427, 539–540.

Tunstall, E. (2013). Decolonizing design innovation: Design anthropology, critical anthropology and indigenous knowledge. In W. Gunn, T. Otto, & R. C. Smith (Eds.), *Design anthropology: Theory and practice* (pp. 232–250). Bloomsbury.

UNESCO. (2009). *Investing in cultural diversity and intercultural dialogue* (UNESCO World Report). UNESCO.

World Rainforest Movement. (2011, 30 August). Forest carbon project in Paraná, Brazil: Reduction of deforestation and persecution of local communities. *World Rainforest Movement.* Retrieved May 13, 2024, from https://www.wrm.org.uy/bulletin-articles/forest-carbon-project-in-parana-brazil-reduction-of-deforestation-and-persecution-of-local-communities

Yujuico, E., & Gelb, B. D. (2011). Marketing technological innovation to LDCS: Lessons from one laptop per child. *California Management Review*, *53*(2), 50–68.

8

THE ARCHITECTURE OF PEACE

Mardelle McCuskey Shepley

While the intentions of the authors of the previous chapters and essays converge, the content covers such a range of approaches that the reader might be left with questions as to the meanings of design and peace. To bring clarification, this chapter explores potential definitions. Two other objectives are reinforced in this concluding chapter. First, we acknowledge the limitations of the impact of the physical and visual environment on human behavior, and, second, we honor the essential contributions of community participatory research.

What Is Architecture? What Is Design?

Defining architecture or design is complicated due to the multiple applications of the terms. The words *architecture* and *design* can be used to describe the underpinnings of a process. For example, an online search using the keyword "architecture" reveals references to computer architecture (system components and goals), brain architecture (chemical links between neurons), and plant architecture (the organization of the plant body), among other topics. Alternatively, definitions of architecture and design can be limited to the creation of physical products and the products themselves. Embracing these two interpretations, this summative chapter is titled "The Architecture of Peace," where architecture is both a process and product.

The chapter and cameo authors move between these two definitions when discussing the impact of their specializations, which include design collaboration, urban design, landscape architecture, architecture, interior design, industrial design, fashion design, design philosophy, and graphic design. Rather than being cowed by radical collaborations, the authors have welcomed disciplinary entanglements as a means of addressing societal and sustainability challenges, and community building. They lean on the methods of disciplines not commonly

DOI: 10.4324/9781003462019-18

associated with design (e.g., anthropology, theology, sociology, philosophy, and environmental psychology) to substantiate the power of design to provide peace.

What Is Peace?

The contributors did not cling to specific interpretations of peace, and a review of the literature reveals that the definitions are diverse. Royce (2004) defines peace as "as a two-dimensional construct with both objective and subjective measures that must be studied within specific micro to macro contexts" (p. 101), suggesting fluid and variable interpretations. The concept is sufficiently complex to have resulted in the development of a research field in and of itself. Multiple definitions are provided by Malley-Morrison et al. (2013).

Nils Gleditsch and colleagues note that:

> Many scholars advocate a broad concept of peace while others hold that peace research should focus on the traditional agenda of mapping and explaining physical violence between groups and investigating how it can be reduced.
>
> *(2014, p. 155)*

Negative peace, which is associated with war reduction, has been the primary focus (Gleditsch et al., 2014). Reinforcing this conclusion, Dixon et al. (2023) reported that US survey respondents defined peace in the context of religion or world peace and linked to safety in relationship to drugs, crime, immigration, and terrorism.

However, the contributors to this book typically addressed "positive peace" and personal peace. Gleditsch et al. (2014) define positive peace as collaboration or integration and reassures us that "patterns of cooperation, the traditional meaning of positive peace, are still being investigated" (p. 155). The definition of personal peace is associated with a person's mental wellness, however, a world in which the participants have achieved personal balance, is a world where negative peace and positive peace will more likely be achieved.

Design Determinism

While acknowledging the significant impact of the environment on behavior, the authors of this book are not implying a determinist stance. Rather, we are suggesting that the designed environment is a catalyst for or contributor to change. As Channon (2022) points out, design may contribute to well-being, but it is not going to single-handedly change the world. We acknowledge that the role of design is to humbly support the human activities required to achieve personal and interpersonal peace. Per Smith (2022),

> Separately, design and peacebuilding are dynamic practices that use engagement, trust building, communication, iteration and an understanding of

> the context to facilitate positive change. Combined they offer matchless opportunities for the formulation and implementation of transformative responses to dire situations and unjust systems.
>
> *(p. 12)*

However, many design researchers use behavioral science to move them beyond their subjective experience, specifically, the science of environmental psychology. The independent discipline of environmental psychology is a mid-20th century construct (Steg et al., 2012) and linked to a variety of pioneering behaviorists. Some of my first readings on the topic were by E.T. Hall (*Hidden Dimension*, 1966) and Robert Sommer (*Personal Space*, 1969), who dramatically changed my trajectory as an architect by making me aware of the psychological impact of space. Most architects and many laypersons support the premises of environmental psychology, but we acknowledge the boundaries of its impact.

The Critical Role of Community

Community engagement and collaboration are at the core of peace, as sharing perspectives among and between groups and individuals is a pre-requisite to conciliation. Community-based design can be community led or community driven (Serrano, 2020) and is directly related to community advocacy. As an example of the interface between design and community advocacy, Liu (2020), proposes an office of the public architect that employs designers that operate in the same way as public defenders by using design experts to support those that cannot afford building-related technical services.

The importance of community engagement in understanding physical and visual environmental needs is a repeated theme in this book, but it is uncommon in practice. Curious as to how frequently community members were involved in participatory research on the topic of access to urban nature, my colleagues and I conducted an in-depth review of the literature. Our search revealed a scant four papers in which communities were directly engaged with a research project not just as respondents but in an in-depth participatory engagement with protocol development, data collection, and interpretation (Shepley et al., 2022). Clearly, more researchers need to fully engage and fiscally support community participation.

Beauty Will Save Us

Both the chapters and cameo authors express how beauty will save us, typically through the vehicle of socially responsible design. Peace is the ultimate form of beauty, and design and design process seeks to simulate beauty. Beauty refers to the beauty within us and around us.

In the physical world, beauty can be expressed in all scales. With regard to buildings, a visit to Corbusier's Ronchamp is predictably lovely and peace engendering (see Figures 8.1 and 8.2). In the context of interior design, a

FIGURES 8.1 AND 8.2 Chapel at Ronchamp, Notre-Dame du Haut, France; Architect: Pierre Jeannette Corbusier.

Sources: "La façade sud de Notre-Dame du Haut (Ronchamp)" (www.flickr.com/photos/72746018@N00/28935542392) by Jean-Pierre Dalbéra (2016) (www.flickr.com/photos/dalbera/) is licensed under CC BY 2.0 (https://creativecommons.org/licenses/by/2.0/?ref=openverse)(left image); "Le Corbusier-Notre Dame du Haut, Ronchamp, 1954" (www.flickr.com/photos/16707687@N00/2501817294) by Rory Hyde (www.flickr.com/photos/roryrory/) is licensed under CC BY-SA 2.0 (https://creativecommons.org/licenses/by-sa/2.0/)(right image).

FIGURE 8.3 Bicycle bell.

FIGURE 8.4 "Lost Cat" graffiti, Artist: John Gavin Mahoney (1968–2023).

library niche can draw you into a similar state of reverie. On the scale of the clothing, Walé Oyéjidé, in his presentation "Beauty Is the Weapon of the Future," described how he used fashion to positively transform both the perception of others and the perception of oneself (Oyéjidé, 2023). On the scale of objects, a fountain pen that fluidly fits into one's hand; a bicycle bell reminiscent of childhood (Figure 8.3), or an aesthetic and thoughtfully designed water bottle can make one smile.

The two-dimensional visual environment can also support peace and beauty. One evening in 2016, while walking across a vacant lot, I (actually) stumbled across the "Lost Cat" graffiti shown in Figure 8.4. Stylistically, it was clear that the sketch, which implied a child-like quality, had been done by an artist. Over the next seven years I spotted "Lost Cat" graffiti in various discrete locations around our town – under the girders of small bridges, on wood pallets at construction sites, and on furniture that was left out to be discarded. The artist, whose identity was publicly undisclosed until his death in 2023, was known by the underground art community and likely contributed to its connectedness. I found his work to be grounding, accessible, and charming, and, in some small way, it enhanced my quality of life each time I encountered it. All of these three and two-dimensional interventions reflect "firmitatis, utilitatis, venustatis" (firmness, commodity, and delight), the terms that Vitruvius Pillio used to describe good architecture.

I frequently find myself in an environment which offers profound peace. Typically, these settings have been in nature – lying in the sun by a spring-fed lake while lost in a forest, crossing a river on horseback, sailing into the harbor after a storm, riding a bike in a rainstorm. All these experiences reside in tandem with a challenge, as harmony is never more obvious than when it is juxtaposed with a bit of adrenalin to stimulate the senses. The challenge of life provides the contrast that enables the experience of beauty.

Returning to the premise of this book, Dostoyevsky's full passage reads "The world will be saved by beauty, and what is more beautiful than love" (Finlay, 2020). The process and product implications of beauty as expressions of peace can be interpreted broadly. Design can play a role by addressing human need, by operating in a transdisciplinary mode, and by sharing knowledge.

References

Channon, B. (2022). *The happy design toolkit: Architecture for better mental wellbeing.* Routledge. https://doi.org/10.4324/9781003277897

Dalbéra, J. P. (2016). La façade sud de Notre-Dame du Haut (Ronchamp) [Photograph]. *Flickr.* https://www.flickr.com/photos/72746018@N00/28935542392

Dixon, P., Lerman, A. E., Vera-Adriazén, F., & Levy, N. (2023, September 20). Americans do talk about peace – just not the same way people do in other countries. *The Conversation.* https://theconversation.com/americans-do-talk-about-peace-just-not-the-same-way-people-do-in-other-countries-213330

Finlay, S. (2020, July 30). Book review: The world will be saved by beauty. *Independent Catholic News.* https://www.indcatholicnews.com/news/40135

Gleditsch, N. P., Nordkvelle, J., & Strand, H. (2014). Peace research–Just the study of war? *Journal of Peace Research, 51*(2), 145–158. https://doi.org/10.1177/0022343313514074

Hall, E. T. (1966). *The hidden dimension.* Anchor.

Hyde, R. (2008). Le Corbusier-Notre Dame du Haut, Ronchamp, 1954 [Photograph]. *Flickr.* https://www.flickr.com/photos/16707687@N00/2501817294

Liu, A. (2020). Toward an office of the public architect. *Log, 48*, 39–52.

Malley-Morrison, K., Mercurio, A., & Twose, G. (Eds.). (2013). *International handbook of peace and reconciliation.* Springer.

Oyéjidé, W. (2023, April 20). *Beauty is the weapon of the future.* Lecture and Cornell University.

Royce, A. (2004). A definition of peace. *Peace and Conflict, 10*(2), 101–116. https://doi.org/10.1207/s15327949pac1002_2

Serrano, P. A. (2020). Designing for with. *Log, 48*, 75–80.

Smith, C. E. (2022). Designing the future now. In C. E. Smith (Ed.), *Designing peace: Building a better future now* (pp. 12–23). Cooper Hewitt.

Shepley, M. M., Ames, R. L., Leitao, R. M., & Coleman, G. (2022). Community-based participatory research on the impact of greenspace on violent crime. *Environmental Research: Health, 1*(2), 022001. http://doi.org/10.1088/2752-5309/aca2d2

Sommer, R. (1969). *Personal space: The behavioral basis of design.* Prentice-Hall.

Steg, L., van den Berg, A. E., & de Groot, J. I. M. (2012). Environmental psychology: History, scope and methods. In E. M. Steg, A. E. van den Berg, & J. de Groot (Eds.), *Environmental psychology: An introduction* (pp. 1–12). Wiley-Blackwell.

Related Bibliography

Brunswik, E. (1952). *The conceptual framework of psychology.* University of Chicago Press.

Dostoyevsky, F. (1869). *The idiot.* CraigsClassics (in English) (Original work published 1913).

Lewin, K. (1936). *Principles of topological psychology.* Martino Fine Books.

Pollio, V. (1914). *Vitruvius, the ten books on architecture* (13–15 BC). Harvard University Press (in English).

Skinner, B. F. (1965). *Science and human behavior.* Simon & Schuster.

INDEX

Note: Page numbers in *italic* indicate figures and those in bold indicate tables.